SOLDIER OF FORTUNE
GUIDE TO SURVIVING THE APOCALYPSE

SOLDIER OF FORTUNE
GUIDE TO SURVIVING THE APOCALYPSE

The Ultimate Guide to Protecting Your Family
Against Societal Collapse

N. E. MacDOUGALD

Skyhorse Publishing

Skyhorse Publishing books may be purchased in bulk at special discounts for sales promotion, corporate gifts, fund-raising, or educational purposes. Special editions can also be created to specifications. For details, contact the Special Sales Department, Skyhorse Publishing, 307 West 36th Street, 11th Floor, New York, NY 10018 or info@skyhorsepublishing.com.

Skyhorse® and Skyhorse Publishing® are registered trademarks of Skyhorse Publishing, Inc.®, a Delaware corporation.

Visit our website at www.skyhorsepublishing.com.

10 9 8 7 6 5 4 3 2 1

Publisher's Note: Some of the material in this book first appeared as articles in *The New Pioneer Magazine*, a Harris Publication.

Library of Congress Cataloging-in-Publication Data is available on file.
ISBN: 978-1-62087-098-3

Printed in China

This book would not have been written without the support and encouragement of my wife, Lara. I am indebted to her for reading my many drafts, running the household while I was holed up, and her trenchant comments and recommendations.

CONTENTS »

PREFACE »

From one survivor to another:

Coming to, I realized my hands and feet were tied to the bed, something was in my throat, and I was drugged up to my hairline. The hospital room was new looking, lots of glass. The thing in my throat had to go. The last thing I remember was walking into the emergency room with my wife.

As my head cleared, a nurse came in, untied my hands, and pulled the long tube out from my throat. I tried to say thank you but all that came out was a croak. My wife explained that it was the next day, and I'd had an action-packed night. I'd had a massive heart attack and died on the exam table. Someone in the ER used the defibrillator paddles on me, but after being shocked I still didn't have a pulse. My wife became terrified. Seconds later a stronger charge restored my heartbeat.

Lying in cardiac intensive care, I thought about what mattered . . . what really mattered. Having received a second chance at life, I wanted to give back. My life of travel and adventure has taught me a great deal, and now I want to share it with you. Why you? Because by opening this book, you're expressing interest in becoming more self-reliant and protecting your family from what's almost upon us.

May this book start you on the path to discovering survival's many facets and providing some of the knowledge and skills you'll need navigating through a changed world.

N. E. M.
Denver, 2012

INTRODUCTION »

Both booties of my dry-suit had leaked so I couldn't feel my feet, hands, or face.
— Sean Cary, CWO3 (SEAL), U.S. Navy (Ret.)

Cary on anti-piracy duty in foreign waters in 2011.

During winter warfare training in February 1986, with SEAL Team One, I was on a combat diving mission in Seward Bay, Alaska. My eight-man squad had launched our rubber raiding craft a few hours after dark. Freezing rain, sleet, and snowfall masked our three-hour surface transit to the target. When the rain and snow touched the bay's saltwater, it froze, creating a thin crust. Our outboard motors were barely powerful enough to drive the boats forward, over, and through the ice. Heavily laden with explosives, machine guns, fuel, and ammunition, we continued through the night, navigating by compass bearing, speed, and dead reckoning.

Every few seconds our small craft's bow would slide up onto the ice, and when most of the boat's weight was on the ice it would crack and break. Like a miniature icebreaker, our lead boat would surge and crack, surge and crack. I couldn't feel my hands. The coxswain had to help me don the three-finger claw mitts over my neoprene gloves. The four of us entered the water—two pairs of divers—cracking and crunching the ice around the boats. Linked by buddy lines and non-verbal signals that enabled us to communicate underwater, we slowly adjusted to the colder cold.

"Protected" from the icy wetness in dry-suits, our hands and heads were covered only with neoprene gloves and a hood. I distinctly remember feeling like I'd been hit in the head with a sledgehammer when going under. It felt like thousands of razor sharp needles were poking my lips and the exposed portions of my face. After gaining control of our breathing, my swim buddy and I collected our composure, gave the OK signal, and descended. We drifted to the appropriate transit depth, squeezed the OK, and kicked off toward the target: A 100-foot cutter moored 1,000 yards away.

About eighteen minutes into the inky blackness, my swim buddy yanked the buddy line and grabbed my bicep. I could hear him screaming underwater for me to make an emergency ascent. I'd been totally absorbed by staying at the correct depth, maintaining speed, and following the bearing to the target. I took positive hold of him, and we ascended slowly to avoid getting an arterial gas embolism.

My buddy's closed circuit (no bubbles) dive rig had flooded. When salt water mixes with the chemical used to scrub the CO_2 out of our exhalations so we can re-breathe them, it creates a poisonous slushy gas called a caustic cocktail. Only breathing fresh air can keep a diver's lungs, trachea, and esophagus from burning and even then, it is not a guarantee. Divers can die if they do not get fresh air within a few seconds of discovering a caustic cocktail.

We reached the surface to find ice blocking our path to fresh air. I remember seeing the glow from the ship's lights 500 yards away. I tried to break a hole in the ice with my fist to no avail. In desperation I used my compass board to break through, and we were able to enlarge the hole so my buddy and I could surface. With a desperate exhalation and inhalation of fresh surface air, my buddy, coughing and choking, brought himself under control. He gargled frigid salt water to rinse the poison from his mouth.

Luckily we were still far enough away that we were invisible to the watchful sentries on the target vessel. Now what? We couldn't go back down, so we had to return on the surface to the rubber boat and our waiting teammates. The only way to move back through the ice was to kick as hard as we could, get an elbow onto the ice, crack and break a piece off, and then do it again. At a rate of two-to-four exhausting feet a minute, we made our painful, freezing way back to the boat. We were hallucinating as they pulled us out of the water.

Both booties of my dry-suit had leaked so I couldn't feel my feet, hands, or face. I was no longer aware of or concerned about the cold. Our teammates stripped off our dive equipment and moved us to shore as quickly as the thickening ice would allow. I recall thinking, *This is how I die,* and, *There are worse places to die than Alaska.* I was delirious when our team moved us into a crab shack and fired up all the cooking equipment. The "black tunnel" was closing in tighter and tighter when a voice deep within said, *You're only twenty-one years old. It's not time to die. There is so much more you want to do. Hang on!*

In the fetal position I felt the pain of rewarming kick in, and the cold came back with a vengeance. My body was jackhammering, and I thought my teeth would break from chattering. Eventually I warmed up and the ordeal ended. We were alive. It was not my first or last near-death experience, but it is by far one of the most memorable. I wanted to live and I didn't give up. The fine membrane between letting go and fighting to stay alive seemed so thin and fragile at the time. Without our teammates to recover us, we would have died of hypothermia and been found the next morning locked in the ice.

Having survived multiple combat deployments during my twenty-five-year career as a U.S. Navy SEAL, severe experiences have perfected my appreciation for survival. Without faith, I certainly would have fallen many times. Reading this book definitely will increase your odds of survival, whether this is your first step in preparing or you've been a survival enthusiast for most of your life.

WHAT EXACTLY ARE WE PREPARING FOR?

Only an intellectual would believe that; the common people are not so stupid.

—Hillaire Belloc

There's something happening here, what it is ain't exactly clear...

—Buffalo Springfield

What exactly are we preparing for? Each of the following chapters could be a book, but you don't have time to read more and there's no time to write more, as critical events are converging at ever-increasing speed. We feel it in our guts. The result is increasing interest in self-reliance and survival products, books, and guns—sales since 2008 have broken records. People refer to ammunition as the new precious metal.

Something's Happening

We're preparing for a reckoning of some kind. Our once-vibrant and dynamic economy is circling the drain. A few countries in the European Union are bordering on bankruptcy. Even the Chinese economy is slowing. When will Israel launch an airstrike on Iran? If Iran develops a nuclear bomb, will it use the bomb on the United States? More people are asking tough questions about the economic future of the United States: Will it be an inflationary or deflationary collapse? Unemployment, welfare programs, and bankruptcies are soaring. Real estate values, savings, and consumer confidence are plummeting. Gold and silver are in short supply, and the prices keep rising. An elite political class—mostly lawyers—have replaced citizen participants in Congress. This class seems more concerned with reelection than with saving the country.

Voters are so polarized that relatively few unaffiliated voters separate the unyielding left and right. The concept of "Too Big To Fail" is being questioned. The Wall Street-White House connection is being criticized. Derivatives and complex credit default swaps, while not fully understood by us common people, are troubling. Is quantitative easing smart in the long run? Is it Constitutional? The government continues spending more money on public education, yet our global ranking for education keeps falling. What is the appropriate balance between individual freedom and privacy and the state's need to spy on terrorists? Should the size of the federal government increase or decrease? The have-nots are close to outnumbering the haves. What then? The social, political, and economic fabric of America is unraveling, and we're watching it happen.

The 2011 and 2012 rioting and protests in the Middle East were partly a result of soaring food prices and unemployment, and political extremism by easily influenced young men. In August 2012, according to NBC News, southern Andalusian Union of Workers (SAT) leaders looted two supermarkets in Ecija, Spain, in a Robin Hood effort, giving the food to the unemployed poor. Five were arrested. Street protests have become commonplace in Spain due to government austerity measures, including spending cuts and tax increases. Is this kind of "expropriation" the beginning of a trend or a one-time occurrence?

Trust Your Gut

The common people, trusting our guts, are doing what is prudent. We're tightening our belts, hunkering down, and preparing. I like to think that we're the common people because we exhibit common sense. We are common because we're not wealthy and have no lobbyists working behind the scenes on our behalf. We simply want to take care of our families. We watched the government's floundering efforts in 2005 when Hurricane Katrina killed more than 1,800 people and left thousands more homeless. We concluded that it's wisest to depend on ourselves. It's not just the United States that does not deal well with major disasters. For example, the Japanese government did a poor job of assisting its citizens in the 1995 Kobe earthquake and the 2011 tsunami from which the island nation is still trying to recover. Or consider the 2004 Indian Ocean tsunami that, according to the U. N., killed more than 220,000 people.

Natural disasters are not the only ones that we're concerned about. Our lifeblood—the dollar—is losing value due to inflation, which means we're paying more for goods and services. What happens if the poor can't afford food and fuel? Will there be rioting and looting? What if terrorists attack using stolen nukes or a low-tech dirty bomb in a dense downtown area? We need accurate and timely information so that we can understand and prepare before the next bubble bursts.

Demanding Situations Make Critical Thinking Mandatory

When confronted with a survival situation, you may first have to deal with emotions that interfere with your ability to think clearly. If your body has begun releasing adrenalin, you can harness it to run, fight, or swim a water barrier. But if you need to size up your situation, figure out priorities, and make a plan, adrenalin or shock will interfere with clear thinking. Take the time you need to regain your composure.

A situation's enormity may overwhelm you. (Imagine you're on a train in a developing country when terrorists stop it. They board and take everyone's valuables and passports. The woman sitting next to you shouts at the nearest terrorist and gets shot dead. She slumps onto you, bloodying your shirt. Minutes later the terrorists leave, shouting and firing weapons in the air. You're in a bad situation, but you're alive. As the train resumes its journey, you can take the time to process what just happened and prioritize your actions.) By breaking down a situation into small steps you can deal with it more effectively.

Taking Small Steps Provides a Sense of Control

Given that none of us can predict how a crisis or disaster may unfold, preparation requires us to think through a variety of scenarios and prepare accordingly. The following organizational approaches to preparation and self-reliance are general enough to be suitable for many contingencies and may assist you in achieving your goals and objectives.

How to Decentralize

Life has become too complex to buy a bunch of products that you think—or a so-called authority thinks—will save your bacon if things get ugly. We must innovate to become more efficient. By decentralizing assets and storing items in modules, we improve our odds of succeeding regardless of which scenario rears its ugly head.

Most of us have one house and a vehicle or two. This situation can limit our thinking because we tend to keep possessions and information in one place. We can benefit by learning about decentralization and how military strategists plan. Versatility is key for military planners, so let's compare our planning to theirs. For example, we keep:

- A full pantry in case of extreme weather or power outages
- Savings in the bank for emergencies
- Guns in our homes to protect our families.

A full pantry is reassuring and convenient. However, storing part of it elsewhere provides more options. A stock of food at a trusted friend's or neighbor's house improves your situation if a fire or extreme weather destroys your home. If you live in a remote area, consider storing nonperishable food in underground caches or a storage facility. If you choose to bury food, be sure to hide a shovel nearby. Rotating your food becomes more problematic, so keep a log noting location, date, item, quantity, and shelf life. Dehydrated food has more nutritional value than canned food because the canning process requires heat, which lowers nutritional value. And consider adding a change of clothes and information (maps) that you anticipate needing.

Savings in a bank account can also be reassuring, but may not be available depending on the scenario. For example, since the fall of 2008, customers attempting to make large withdrawals from certain banks and stock brokerages were informed of a limit on the amount they could withdraw. As a result, many savvy customers began keeping more cash at home. Also, when smaller banks go bankrupt, bank regulators can declare a "bank holiday" until a stronger financial institution acquires the defunct one. This process can take several days. The impact of an interruption of financial services can range from inconvenient to catastrophic. Carrying extra cash in your wallet or vehicle also is prudent. If you travel internationally, be sure to carry all the currencies of the countries you will travel through.

A gun in your home is logical because when needed, it's needed instantly. The key here is to keep *only* the gun or guns needed for defense. Firearms in a locked safe are of little use during an emergency, like a home invasion. By storing some weapons with a friend, you may accomplish more than one goal: You decentralize valuable assets, and if the friend has no weapons, he or she can use yours for protection.

Decentralizing requires forethought and discipline but can yield lifesaving benefits and peace of mind in an uncertain future.

How to Modularize Your Preparations

If you're considering decentralizing your assets and resources, why not divide them into logical groups as well? For example, modularizing your pantry contents, ammunition, and tools enhances convenience and portability.

Pantries are both convenient and useful in case of a disaster. Most rural and mountain dwellers keep a supply of canned goods just in case. But what if the food has to be moved quickly? The answer is that it would take hours, assuming you could find boxes or bins to contain the food.

Pantry Management

Milk crates facilitate pantry management by keeping similar items in plastic milk crates. Milk crates have three advantages: They're strong, light, and cheap. Used milk crates can be found in flea markets, garage sales, and junk stores for a buck or two. It's better to get the crates now and fill them with pantry items than to attempt filling the crates at the eleventh hour when every minute counts. Before and after photos of my pantry demonstrate the organization milk crates provide. Not only can you move items quickly, but they're already grouped according to your organizing principles.

Pantry foods are good quality, organized, and convenient. But can they be moved quickly if necessary?

Milk crates group foods into modules on pantry's left side. One shelf was removed to accommodate the crates.

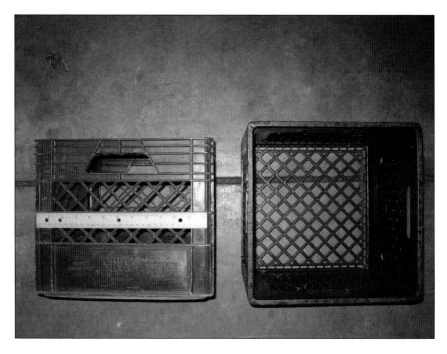

Inexpensive and found everywhere, milk crates are strong and light. They help to organize and modularize smaller items. Ruler shows scale.

Ammunition

An urban-dwelling acquaintance of mine owns several thousand rounds of rifle ammo in case of civil unrest. He discards the cardboard cartridge boxes and keeps his ammo loose in military ammo boxes so he can access it quickly. The waterproof steel boxes are inexpensive and nearly indestructible, so he thinks his system is a good one. And it is, as far as it goes. He has obtained the needed cartridges, but they are not in useful modules. I ask him how many magazines he owns for his big rifle. His answer is fairly common: Four. I then ask him how he plans to reload his magazines during a stressful situation. His confident smile falters for a second, and he says that he understands what I'm driving at. When I last saw him, he was ordering more magazines. By keeping cartridges in magazines—and rotating them if you are concerned that the springs will weaken if kept compressed over time—you have the ability to load immediately.

When seconds count, loose cartridges are less useful than those in a magazine. If you're concerned about damaging magazine springs by leaving them compressed, rotate the magazines.

Tools and Spare Parts

When repairing things around the house, most of us grab the tools needed for the job (carpentry, mechanical, electrical, plumbing, or gardening), put them in something we can carry, and head in the problem's direction. Having owned several properties, I developed a modular method: A separate toolbox for each trade so that I can grab one and go. This modular method saves time and also works for hobbies and sports: A container for each activity ensures that you won't forget something.

Left: Red metal box contains assortment of washers; red tray holds socket set; plastic box contains plumbing tools, adhesive and supplies. Right: Grey metal toolbox and red tray contain carpentry hand tools; wooden tray holds electrical parts and tools; metal tray is full of files.

Assembling a Bug-Out Kit

If you're looking for a brief list of things to buy, read no further. A bug-out kit is like a house—it means different things to different people. So what follows isn't primarily about equipment: It's about the thinking process behind the contents of your kit. In Chapter Eleven, you'll discover several recommendations for items that you'll need to survive any scenario, but first an examination of how you can determine what goes into your kit.

DIFFERENT SCENARIOS REQUIRE DIFFERENT APPROACHES

A bug-out kit to flee an urban riot is different from the one used to escape an impending economic collapse, a weeklong power failure, or a wildfire. What you foresee determines which items you'll include in your kit. An appropriate kit may be a small belt pack containing cash, a few gold coins, a couple of credit cards, and a passport. . . . Or it may be an oceangoing sailboat with adequate supplies to keep six people clothed and fed for months.

A bug-out kit can fit in your pocket or be a bulging backpack or loaded recreational vehicle.

Because your kit may save you and your loved ones, plan for likely scenarios, which will shape your list of supplies:

- Are you living in a large city planning to escape to a rural retreat 100 miles away?
- Do you plan to stay in your urban unit's safe-room undetected by intruders and looters?
- Is your frame home surrounded by forest with the ever-present danger of wildfire?

- Do you plan to travel to an ally's house four miles away where you have stockpiled a cache?
- How long will you be gone? Days? Weeks? Months? Years?

Even the ideal kit is useless if you can't get to it, which raises another question: Do you need multiple kits? Numerous kits provide numerous options, which is the real purpose of a bug-out plan. Keeping kits in various locations increases the likelihood of success for your plan. Flexibility is key to preparedness.

A PROVEN PLANNING METHOD

Planning backwards from a desired outcome is a method that military logistics professionals use. It is especially useful when working to a deadline or needing to determine a budget for your outcome. Below is a scenario and desired outcome—with specific conditions and restrictions—to help illustrate typical steps in the reasoning process. As a safety factor, military planners often request 20 to 25 percent more supplies than needed.

The Scenario:
Jim and his wife live in a flood-prone area beneath an earthen dam. The size and shape of the flood zone is known from previous floods. In the past, as the water level increases, authorities begin broadcasting evacuation alerts that provide an average of twelve hours warning.

Restriction:
Jim's wife is confined to a wheelchair because of an accident. Good questions include: Where will Jim be when the scenario becomes reality? Is he at home? At work? Traveling on business overseas? For this scenario Jim does not travel for business and works in town eight miles from home.

Jim's Kit and Plan of Action:

Jim has assembled the items he'll take and has containers stored exclusively for them. He has practiced packing the listed items and knows that it takes a little more than one hour. Hiking out or riding a bicycle is out of the question because of his wife's condition. Jim will use his primary vehicle, a four-wheel-drive van with a lifting apparatus for the wheelchair. Is what he's currently driving suitable? Does it have adequate ground clearance for overcoming curbs? Yes. Can it push obstacles out of the way and carry him safely out of the flood zone? Yes. The van contains tools, an extra spare tire (not an undersized temporary), spare parts, and a heavy-duty jack. Because he needs to drive only eight miles to reach high ground, he need not cache fuel along the escape route. But he needs to make sure he always keeps the fuel tank one quarter full. What if his primary escape route is impassible? Does he have a secondary? Yes. What if Jim's vehicle does not start? He plans to put his wife in his motorcycle's sidecar and take only photos and documents with him. If the motorcycle does not start, he's made mutual-aid arrangements with neighbors who have agreed to stop by his house before bugging out.

Evaluating the preceding bug-out kit and action plan is relatively straightforward: Jim has anticipated numerous contingencies and has plans and backup plans that will likely succeed. He has generated specific questions and answered them in a methodical and thoughtful way. If you plan to travel with others, ask them to develop questions as well. The more problems you can anticipate, the less psychologically off balance you will be if they occur. A vague idea of throwing a few items in your vehicle and driving fast is not a kit or a plan—it's a knee-jerk reaction that does not stand up to examination.

The following emergency checklist categories are not intended to be comprehensive. They are placeholders to stimulate thinking about creating relevant categories for your checklist.

Emergency checklist				
	Destination	Evac. routes/ vehicles/ convoy	Rendezvous point(s)	Kit contents
Scenario 1, pandemic or _____				
Scenario 2, earthquake or _____				
Scenario 3, civil unrest or _____				

Budget

While you can prepare a useful kit on almost any budget, the bigger the budget, the more options you have. For example, imagine that a virulent and deadly pandemic is announced. You decide to take your family to live isolated in the wilderness for a few weeks to wait it out. A low budget will enable a tent and supplies. A high-budget kit might consist of a recreational vehicle (RV) or a remote cabin stocked with every need and want. Either budget will provide the isolation you're seeking.

First Needs

Whatever your plan, you'll need drinking water. If you have access to fresh water along the route, you can turn creek water into drinking water by filtering it or adding bleach or iodine (the amount to add depends on how murky the water is)—if not, you'll have to carry it. Food is essential if you're hiking or biking out, especially in cold weather. If you plan to travel by motor vehicle, it also may require water and antifreeze for the cooling system. You probably won't have time to prepare food on the way so buy convenient, balanced nutrition that you can eat on the run.

Clothing

Suitable clothes can protect and maintain your body temperature. You don't need a wardrobe, but you need tough clothes and footwear to get where you're going. Avoid cotton, which stays wet and can cause hypothermia in cold temperatures. Choose synthetics or wool, or a blend of the two . . . and choose broken-in boots rather than new ones. Make sure you do not forget a hat, which provides protection from sun and cold. See Chapter Eleven for detailed information about winter clothing.

Shelter

Shelter is crucial if you're traveling in extreme temperatures. The tent, RV, or camper you have certainly beats hoping to find accommodations on the road. Choose a quality, serviceable tent that is easy to erect while wearing gloves or mittens, and preferably one that you can erect while in it—this keeps you out of inclement weather. See Chapters Seven and Eleven for more detailed information about tents and tarps. If you're considering buying a camper or RV, plan for sufficient storage for survival items.

Vehicles

If you plan to use a bicycle to get from your suburban home to your staged vehicle—but don't plan to use it afterward, buy an inexpensive one. If you are buying a motor vehicle to take you and yours to a remote retreat, then you probably want to invest in a reliable unit with all the items you'll need in such an important vehicle. You'll want a robust vehicle that can go cross-country if need be. It will benefit from reinforced bumpers to ram or push an obstacle out of the way. Tough tires with sufficient tread, and a full-size spare tire can mean the difference between escape and being stranded.

Supplies

If you're considering a convoy, you'll want dependable two-way communication. And you'll want spare parts—headlights, for example, will break or burn out at the worst time—and tools to install them, as well as ample water and food for your journey. You'll need a first aid kit, extra containers of fuel for vehicle and camp stove, maps and compass (or a GPS system and data), camping gear, fire extinguishers, and clothes to match the season. You may also need weapons to defend your family. If you remember nothing else from this chapter, *don't* buy a gun and ammunition hoping to learn to use them after you bug-out. Buying a piano does not make someone a pianist: Only regular practice over time makes one competent with a piano or firearm.

Making a List of Useful Items

While you have time, create a customized list of what you'll need to take so that when the time comes, you'll be ready to act. Reviewing this chapter will provide a good starting place for your list. Start with water and work your way through less critical items. Consider the next season when choosing clothes.

Worst Case

In the event that you become separated from your bug-out kit or cannot get to it, you may feel especially vulnerable. If you haven't already, read *Deep Survival* by Laurence Gonzales. The book will help you to understand or remember that even when stripped of vital equipment you've still got what it takes to get you and yours through an ordeal.

2

YOUR BRAIN AS THE ULTIMATE SURVIVAL TOOL

The fine spread of antlers was locked fast in the tangled roots of a fallen tree. Kolemenos . . . swung the shining axe blade down with a vicious swish . . . and the deer slumped . . . Paluchowicz busied himself gathering wood, laying and lighting a fire, while the rest of us built a shelter and finished the butchering.

—Savomir Rawicz, *The Long Walk*

A human being should be able to change a diaper, plan an invasion, butcher a hog, conn a ship . . . set a bone, comfort the dying, take orders, give orders, cooperate, act alone, solve equations, analyze a new problem, pitch manure . . . fight efficiently and die gallantly. Specialization is for insects.

—Robert Heinlein, *Time Enough For Love*

The Ultimate Survival Tool

Killing, skinning, and butchering a trapped deer in the wild after escaping a Soviet labor camp in Siberia is something you probably won't have to do. But your attitude toward solving problems, especially by improvising, can help you to overcome a life-threatening situation. Your brain—its knowledge and skills—is the ultimate survival tool. You may have little experience improvising to solve practical problems. (*Practical* is the key word in the previous sentence: Most of us solve problems daily.) This chapter provides general and specific ways to acquire useful new skills, as well as examples that can make you more self-reliant. A can-do attitude will help you in any stressful situation.

Consider the following as you investigate ways to educate yourself about self-reliance. You can discover what works from those who solve problems for a living. For example, farmers and ranchers use what they have at hand to fix broken equipment, mend fences, build structures, maintain vehicles and implements, and so on. They're practical people who often solve more real-world problems in a day than most of us solve in a week. If you know farmers or ranchers, volunteer to help them so that you can observe and learn. You might find that keeping a dog in front of and behind you is wise in cougar and bear country. Yes, I have two dogs. Or you might discover that putting your knee into a horse's ribs as you cinch the saddle girth keeps the horse from inhaling and leaving the girth loose.

Survival Prospects for Specialists and Generalists

Specialists usually have tunnel vision for one facet of a large and dynamic system. Specialization has enabled people to make breathtaking scientific, medical, and technological breakthroughs during the last century. Specialization also is crucial for those who desire to make large amounts of money in business. The problem is that most specialists don't thrive in survival situations. The exceptions are specialists with skills that are in demand during major disasters: physicians, nurses, paramedics, firefighters, and the like.

Generalists usually see the system as a whole and tend to perform better in adverse situations. The longer a disaster's duration, the better off generalists will be. For example, a general contractor or shade tree mechanic will likely find ways to earn a living while a salesman, retail worker, and artist may have trouble making ends meet.

Relevant Activities vs. Passive Entertainment

Doing relevant activities like swimming, running, hiking, backpacking, rock climbing, mountaineering, cross-country and downhill skiing, archery, fishing, hunting and other shooting sports, martial arts, and camping can improve your odds if you are thrust into a scenario where you can use any of these skills to get out of trouble. Television

An afternoon well spent in Eldorado Canyon, Boulder, Colorado.

is not useful for training unless you are viewing a how-to video teaching a skill best suited to that medium. For example, watching a video demonstrating skinning and gutting an animal certainly beats reading about it. Likewise, a video showing how to clean and suture a laceration is better than reading about it and looking at illustrations or still photos. YouTube.com contains a great deal of useful survival information, and I recommend it for learning complex skills. That said, no video can replace hands-on experience.

Emergency Medicine

Receiving emergency medical training is mandatory for anyone spending time away from civilization. Children get skinned up frequently, so you need to be able to make informed decisions about how to treat them, especially when the nearest emergency room is hours (and possibly days) away. Wilderness first aid and wilderness first responder courses are a great beginning, and becoming a Wilderness Emergency Medical Technician prepares you to treat a wider range of diseases and injuries. When looking for a training facility, make sure that the instructor for the hands-on (clinical) portion is someone with years of experience in the wild; not someone who worked out of an ambulance or emergency room. Regardless of the scenario, knowing emergency medicine is a valuable skill.

Wilderness as a Teacher

Being in the wilderness forces you to improvise, as you're not able to get to a store and buy a forgotten item or tool to fix what's broken. Car camping, backpacking, rafting, and boating are great ways to stretch yourself to solve problems. When your camp stove refuses to light, can you troubleshoot the problem and use the pin on your pin-on compass to clear a clogged orifice? Can you start a fire by using damp matches or no matches at all? If you lose a paddle, can you make another by attaching the top from a cook pot to a branch? Are you skilled enough using a map and compass to revert to them when your fancy GPS

Author leading moderate technical rock climb in Colorado backcountry.

gizmo fails? Can you immobilize a broken tibia in the wilderness by using branches as splints and clothing to fasten them? Wilderness forces you to make do, to think creatively, to adapt, and to overcome.

Organizations like the Wildlife Awareness and Education Institute (WAEI, waei.org) in Colorado, and Coyote Trails School of Nature (coyotetrails.org) in Oregon exist to teach novices the skills to become outdoorsmen. Look around for hiking and mountaineering clubs or, if you're an accomplished mountaineer, join a search and rescue organization. Headquartered in Wyoming, National Outdoor Leadership School (nols.edu) offers numerous courses in North America and worldwide, and enjoys a solid reputation. NOLS also offers degree programs.

Volunteering

If you're not good with tools, become proficient. Otherwise they will do you little good. One useful way to learn practical skills is to volunteer for charitable organizations that build homes for the poor. You can arrive unskilled in the morning, and by day's end you'll know a couple of fundamental skills to use as a basis for more advanced activities.

Joining a volunteer fire department is an ideal way to learn useful skills and work with those performing a critical community service. Late in life I joined a volunteer department in my tiny mountain community. Those four years abounded with learning and relearning. Working side by side with the experienced and inexperienced from different backgrounds reinforced my people skills and patience. Learning the numerous pieces of specialized equipment, vehicles, and techniques was challenging. Becoming a First Responder updated the many first aid and CPR classes I'd taken over the decades.

Fighting wildfires in our heavily forested village was the practical result of all our hands-on training and fitness workouts. We saved a neighbor's house, and his heartfelt gratitude was more than ample compensation. Most rural fire departments maintain and repair their vehicles and equipment, which was another learning opportunity. And the situational awareness required when training to pry open a jammed door on an unstable rollover vehicle was significant.

You'll need an exam from a physician and a letter attesting that you are healthy enough to perform firefighter duties. Some volunteer departments require that you live within the fire protection district that you are applying to, so do your research.

Try Something Different to Learn New Skills

If you have the opportunity or feel stuck in your current occupation, do something different. In 1991, during a recession, a friend offered to help get me a job in the oilfields of Alaska. At that time I couldn't get work as a writer or editor in Colorado, so I accepted his offer. I got hired into the nation's biggest old boy's club, the North Slope oilfield. I became a warehouseman, a job that I was completely unprepared for . . . and that was its beauty. It forced me to learn new skills and meet new people. Among other things, living and working north of the Arctic Circle in Prudhoe Bay taught me to wear a hat outdoors to prevent an almost immediate headache. I operated a forklift and dealt with every kind of person that walked into my warehouse. The warehouse demanded organization, so misplacing an item meant hunting for it when time mattered. Working eighty-hour weeks taught me a great deal about my limits, and living in an oil-camp for a year reminded me that good people skills are crucial for getting things accomplished. Doing something different can expand your horizons and increase your mental agility to prepare you for the difficult times to come.

Travel as Teacher

Travel is a great way to learn practical skills and improvisation, because you experience how others live and work. While travel to developed nations certainly is better than not traveling, experiences in developing countries are best for learning alternative ways of doing things and seeing how few possessions others have. The following are a few experiences from my travels abroad that demonstrate their benefits as learning opportunities.

My deployment in the U.S. Army was to the Military Assistance Command Vietnam (MACV) in 1966, as the war was ramping up. Learning that I couldn't remain vigilant 24–7 was a valuable lesson. Trying to stay vigilant all the time is counterproductive. I'll use green, orange, and red to help illustrate my point about states of vigilance. Green means having a relatively low degree of situational awareness. When entering a dangerous environment, your vigilance increases to orange. Potentially dangerous places include liquor stores and convenience stores because armed robbers frequently target them. Poorly lighted parking lots and alleys, and neighborhoods notorious for gang activity and high rates of violent crime also demand increased vigilance. If you're walking or driving through a dangerous area and notice that you're being followed, you'd become extremely vigilant and go to red, which could mean altering course, using your cell phone, or unholstering a firearm. In dangerous locations, the alternative to vigilance is victimhood.

In 1970, while a student, I spent five months in Kenya and traveled to Uganda and Tanzania. Volunteering at the Outward Bound Mountain School in Loitokitok, Kenya, I became a temporary instructor. The all-boys school taught orienteering (map and compass work), swimming, and backpacking. While on the graduation climb to the summit of nearby Mt. Kilimanjaro, one of the boys began having symptoms of pulmonary edema at an elevation of about 14,000 feet. Because a boy in the previous class had died under similar circumstances, we had to place him in a litter and descend to the main camp at about 8,000 feet. The journey became an ordeal as we hiked through the night carrying the steel litter containing the boy. Exhausted, we reached the main camp at first light. Whether you choose to take an Outward Bound course or teach one you will learn a great deal.

In 1983, the opportunity arose to travel to Nepal and trek up Mt. Kala Pattar, which was the base camp for expeditions attempting to climb Mt. Everest in the Himalayas. I jumped at the opportunity. Flying into Kathmandu, I realized it was unlike anywhere I'd been. The thick mix of wood smoke and two-cycle engine exhaust was everywhere. The following day, we were driven by truck from the city to a trailhead. After a few days of trekking, one of the porters cut his foot. Porters carry their shoes, when not wearing them on rough terrain, to prolong the life of the soles. The leader halted the trek to examine the man's foot. The Sherpas talked for a few minutes, concluding that I should treat the man's foot. He'd cut through one-quarter inch of callous on the ball of his foot. I cleaned it as best I could. The man didn't even flinch when I poured Mercurochrome into the deep cut. Through the interpreter, I told him to keep it clean knowing it would be almost impossible considering the conditions that we encountered gaining or losing thousands of feet of elevation each day. At the next village, the porter left our party to let his foot heal. I never saw him again, but the experience has remained in my memory all these years. Travel in developing countries exposes a traveler to unusual opportunities and circumstances and is an education in itself.

How to Find Opportunities to Increase Self-Reliance

Becoming more practical and resourceful is a matter of commitment, repetition, and duration. Remember your first time riding a bicycle? It was hard, yet you were able to master it. The same goes for practical skills. Make no mistake, you needn't become a carpenter or mechanic, but you must understand their trades and be able to use their tools. Many bike stores and free universities offer bicycle-repair courses, which is a great way to learn new skills. Assembling furniture and equipment is another way to become more self-reliant while saving money. All do-it-yourself projects push you to find ways to solve problems.

For example, if the stores are closed and you don't have a clamp or vise to hold two pieces of wood together for gluing, how can you proceed? You can twist wire or rope around the pieces to squeeze them together. You can place heavy items on the pieces to hold them together. Do you run your generator dry to prevent varnish buildup in the carburetor? Can you clear a clogged vacuum using a coat hanger? Clear the throat of someone who is choking? How many ways can you configure a tarp to make a shelter? If you own a gun are you positive that you know how to shoot it accurately while under stress? You'd better be absolutely positive, and trust me you will be under stress.

How-to Tips and Techniques

The next time you hire a plumber, carpenter, repairman, or tile setter, instead of leaving the tradesman alone to perform the task, ask if you can watch and ask questions. Observing and taking good notes can enable you to do a similar chore in the future. Shoot a video of a skill that you find challenging. For example, a video of sweat soldering a length of copper pipe (male) to an elbow fitting (female) may be useful, because you'll see the copper turn color as it's heated with a torch and you'll see the flux pull the solder into the joint. When the solder cools, it should look mirror-like (grainy, grey solder is a cold solder joint and must be redone because it will leak under pressure).

One consequence of doing tasks that you previously hired others to do is that you amass tools, which increases your odds of having what you'll need for the next job. You may be surprised at how quickly you master most of these trades. I'm not suggesting that you begin by

Author performing preventive maintenance on a snowblower.

trying to fix your laptop computer: You can start with simple repairs and raise the difficulty as your experience and confidence increase. The broader your skills and ability to improvise, the likelier you are to survive a trying situation.

Teaching Is Learning

Teaching is a fine way to reinforce one of your skills, as you must divide it into learnable steps for the student to build upon. For example, you cannot teach someone to shoot without first teaching the student about safety. You must teach nonshooters to relax and breathe so that they don't shake. The student cannot aim until learning what a proper sight picture is. You also need to demonstrate a solid stance to a nonshooter. A beginner will not become accurate unless you teach them the importance of a controlled trigger press. The steps for safely clearing a jammed firearm must be taught in the

Author taking aim with Model 21 Glock .45 ACP semi-automatic pistol.

correct sequence. . . . And no one can become proficient without practice. After the student becomes proficient, you can begin introducing and teaching tactics. The same instructional design process applies to teaching how to clean a house or build a nuclear reactor. Teaching is learning.

Examples to Stimulate Your Thinking

Imagine that you are driving through a remote stretch of the intermountain west when you get a flat tire. You change the tire, but the hubcap full of lug nuts slips into a deep, fast-flowing creek. What do you do? You can remove one lug nut from each of the other three wheels and use them on the replaced wheel, so that you can drive to a town where you can buy more lug nuts; a simple problem and solution.

Imagine that the elevator you're riding in stops suddenly, and then the light flickers. You try the elevator's controls but get no response. You check to see what you have to work with. A disposable lighter or tiny LED flashlight provides light: Either can enable you to see where the emergency alarm button or telephone is located. A hard-sided briefcase enables you to sit while you wait. A whistle can signal for help. If the elevator light remains on, a book can make an hour seem like a minute. That half-eaten energy bar in your pocket will sustain you better than eating lip salve. And a cell phone trumps all of the preceding items.

Imagine that you're driving through a western state in winter when you become sleepy. You pull off the road to take a nap. That's when you remember not adding antifreeze last fall. You don't want to turn the ignition off because the water in your radiator might freeze, but you can't leave the vehicle running because you may not have enough fuel to reach a gas station. You open the trunk to get your sleeping bag. Grabbing the bag, you take stock: a spade, tennis ball, paperback novel, a plastic bag from a supermarket, two large bottles of vodka for a party, and a can of flat fix. What do you do? The radiator is low on coolant so pour the vodka in, wait a few minutes so the alcohol and water can mix, turn the ignition off, and sleep soundly knowing that the mix of ethyl alcohol and water won't freeze in the climate you're traveling through.

Learning by Observing

Workers have blocked off a street and are felling a large tree. Where does the chainsaw operator make his first and second cut? Why is the third cut on the opposite side of the tree? Does the tree fall in a controlled manner or suddenly? Does the chainsaw operator shout anything after the third cut? Sometimes survival can depend on a detail that you read about or observed years ago—a detail that you thought you'd forgotten until the right situation triggers your memory.

Author after felling a tree and sectioning the trunk on his Colorado mountain property.

Become a lifelong learner to increase your chances of survival. In general, reading how-to books is more useful than reading fiction. Television is useful for getting breaking news about the geopolitical situation and other current events. If the Middle East is at war, you need to know. If the price of wheat is about to double, you need to know. If ammunition is about to become scarce, you need to know. Keeping current on the knowledge and skills you need to be prepared is time intensive and can be expensive, but the alternative is unthinkable.

3

SEVERE WEATHER EVENTS

Sunshine is delicious, rain is refreshing, wind braces us up, snow is exhilarating; there is really no such thing as bad weather, only different kinds of good weather.

—John Ruskin

Evidently, Mr. Ruskin has never experienced a hurricane or tornado. Having grown up on an island in the Atlantic, I've experienced numerous hurricanes—ponderous wet beasts that enable television journalists to leave the studio and appear heroic. Because hurricanes—called typhoons in the Pacific—make the news every summer and fall, I won't dwell on them too much. In 2005, Hurricane Katrina taught a dramatic lesson about how the federal and state government perform during a major disaster. Who can forget the scenes of helpless people trapped on rooftops waiting for overwhelmed rescue personnel. Immediately following the huge storm, many law enforcement personnel were tending to their families instead of reporting for duty. So after wind and rain came lawlessness.

Anyone with a radio, television, Internet connection, smart phone, or neighbor, knows of an approaching tropical storm or hurricane. Residents also know a hurricane's name, category, direction, and estimated time of landfall a few days before it hits. So please understand if I'm less concerned about hurricanes than other threats. Preparation plans for hurricanes can be found online in numerous websites, including the Federal Emergency Management Agency (ready.gov/ hurricanes), the American Red Cross (redcross.org/prepare/disaster/ hurricane), and Florida's storm-battered Broward County (broward. org/HURRICANE/Pages/Default.aspx). The only upside of hurricanes and earthquakes is that their death and destruction have led to stronger building codes.

During the lead-time before a hurricane reaches land, pack up your family and drive toward high ground to avoid flooding. If you must stay home, preparations include: fueling your vehicle(s), plywood to cover glass, fire extinguishers, flashlights, extra batteries, food, and plenty of water. Don't forget prescription medications and eyeglasses, ice to cool perishables, candles or LED lanterns, cell-phone charger for your vehicle, battery-operated radio, and a first-aid kit. Sufficient propane for an outdoor grill is wise and rarely thought of. Electronic garage door openers won't work during a power outage, so either leave the garage door open or test it manually to make sure that you can lift it. Have enough cash on hand to buy supplies and vehicle fuel—most stores won't accept credit cards or checks during a power failure. And spare batteries for kids' toys and games will help to maintain your (and their) mental health.

Tornadoes

Tornadoes Require a Fast Response

Twisters are different and more dangerous than hurricanes, as they can develop suddenly into roaring killers, demolishing everything in their path. They can drive pieces of straw through boards, toss trucks like toys, and lift houses off their foundation. For those living in Tornado Alley, an above- or in-ground storm shelter is a responsible investment to protect your family. Only purpose-built storm shelters can withstand an EF5 tornado. The Enhanced Fujita scale, which the National Weather Service uses, ranges from EF0 (65 to 85 mph) to EF5 (200 to 322 mph). The deadliest tornado in the United States occurred in March of 1925. The tri-state tornado killed 695 people, devastating more than 200 miles in Missouri, Indiana, and Illinois.

Constructed from steel, high-strength plastic, fiberglass, or reinforced concrete, shelters can be viewed and purchased online: Enter "tornado shelters for sale" into a search engine. A variety of prefabricated shelters are available, or you can have one built to suit your needs. Shelters seat from four to forty people and range in price from $3,200 to $20,000 plus excavation fees. Most shelter companies are regional, which means some will deliver and help with installation and some won't, so do your research carefully. Like houses, shelters range from modest to luxurious, so determine your budget before shopping online to avoid buying more shelter than you need. Compare warranties as well as materials, construction, features, price, and build quality. If your storm-shelter will double as a safe room, blast shelter, or fallout shelter, the cost will increase according to your requirements. Other shelter uses include meditation and study, sleeping, practicing musical instruments, and storage for valuables. Virtually all storm shelters have stairs so you will need to carry the disabled up and down so plan accordingly. Features include:

- Enhanced shelters that are covered in three feet of earth
- Installation under concrete garage floors
- Warranties ranging from five to forty years (a few are transferrable, great if you sell your home).

After reviewing dozens of shelter websites I came away impressed with how American ingenuity and enterprise have responded to the need for protection from tornadoes.

How Uncle Sam Assists Homeowners to Get a Shelter

The following are a couple of examples of government funding for storm shelters in the form of loans and financing programs. Block grants to communities and states also are available through FEMA's Hazard Mitigation Grant Program and FEMA's Pre-Disaster Mitigation Funds.

SBA Disaster Loans

A disaster assistance loan from the U.S. Small Business Administration is for homeowners to repair or rebuild a damaged or destroyed home. Borrowers can use a portion of the loan proceeds to construct a safe room. The SBA also can increase an approved disaster loan by up to 20 percent to pay for the additional cost of a safe room.

FHA Mortgage Insured Financing

HUD permits borrowers to include storm shelters as an eligible item for FHA 203(k) rehabilitation loans and FHA 203(b) financed new construction. Shelters financed with FHA-insured mortgages must be built consistent with guidelines in FEMA publication 320 as well as the National Performance Criteria for Tornado Shelters.

When a Tornado Gets Close

Preparing for a tornado is like preparing for an explosion in your home. Given the few seconds you might have before a tornado rips your home into kindling, the most you can do is grab your loved ones and run for your storm shelter. If you don't have a storm shelter and the tornado is howling, then get to a corner in your basement or take cover in a closet or bathroom (preferably without a window).

If you're driving during daylight and suddenly darkness overtakes you, try to determine the tornado's path and drive at a right angle to it. While a twister's rotational speed may exceed 200 mph, the ground speed may be only a few miles-per-hour, so driving fast may not be the safest course of action. If the tornado seems to be following your vehicle, drive as fast as conditions permit and turn left or right when you can—remember that tornadoes churn up dust and debris. If you're driving in open country, look for a tunnel, culvert, or ditch to crawl into. Tornadoes have knocked over boxcars, so a large one can most likely pick up whatever you're driving.

Snowstorms and Blizzards

When the blizzard hit, my wife and I weren't prepared. Snow fell for three grey days during Colorado's blizzard in March of 2003. The weather wizards said we'd get between a foot and eighteen inches, so I didn't give it much thought. I awoke to more than a foot of dense snow, which was coming in sideways. I went out, started the snow-blower, and began working to clear the driveway. My wife phoned from Los Angeles where she was working for the week. I told her we'd gotten some snow but it looked beautiful. After working in the house for a couple of hours, I looked out the window and was surprised to see no sign of my morning labor, as the snow and wind had erased my efforts. I went out, gassed the snowblower, and began again. My car was trapped in the garage, so I needed to clear the driveway. After blowing the driveway and shoveling the walk, I worked in the house for a while and then looked out the window again. There was no sign of my second pass with the snowblower. The heavy snow contin-ued to fall horizontally. I decided to attack the snow in the morning when I'd be fresh.

The morning was no better. More than two feet had fallen and more was predicted. According to the latest weather forecasts, we'd get another foot or so. I was preparing dinner when the power went out. I had no idea then that my power would be out for a week. Lighting both woodstoves, I continued with dinner. The Coleman gas lantern was hissing when my wife called my cell phone. I discussed the storm and she discussed the sunshine. The third dawn was fairly dark. Snow had drifted to about six feet, which meant that my car was useless. I was grateful that my wife's car was in the

Author leaving his mountain property to deliver a pair of snowshoes to a neighbor.

airport parking structure. I saw no sense in shoveling until the snow stopped. I made a large breakfast and stoked the fires. After realizing that I'd have to go for more wood, I strapped on my snowshoes and fumbled to the snow-covered woodpile, making several trips. The dogs could barely get around.

The fourth dawn broke clear and bright. I ate a big breakfast, grabbed the chainsaw, and began clearing our road of downed trees so the snowplows could get through. I had no idea that two of the three five-ton plows had broken down. We'd received more than four feet of tree-breaking snow. I returned home thinking a snowplow would soon come to clear the road, which would make walking without snowshoes possible . . . as I still had no power. A radio newscaster said that Denver International Airport had closed until further notice—my wife had to stay in sunny Los Angles.

Two days later, our road remained impassible. My wife called saying she'd fly in later that day, as the airport was cleared of snow. I told her to meet me at the post office a half-mile from our home. When she asked why, I said I'd explain later. It was no use trying to describe the snowfall. I carried her snowshoes and a pair of boots to the post office parking lot. We trudged uphill toward our house, and she began to see downed trees and destroyed power lines. No vehicles could move on or off our road. She saw food from our freezer sitting in the snow near our front door and said we should snowshoe back to her car and drive down the hill to buy a generator. I agreed and put a sled and rope in her car. We bought one of the last generators in the store and drove back to the post office parking lot. We got the machine on the sled, pulling it with the rope toward our house. After stopping several times to rest, I finally tied the rope around my waist and pulled while my wife, also on snowshoes, helped to push it over the snow covering our steep driveway. It was great to have power again after days without it.

Lessons Learned

We needed more food in our pantry and more stored gasoline for the generator. I built oversized shelves in the basement for a large pantry, stocking it with hundreds of pounds of beans and rice and commercial-sized canned goods. The old cistern became a storehouse for several six-gallon containers of stabilized gasoline and diesel fuel.

Author's leaky cistern stores water, food, and fuel.

Author using drill powered by solar generator assembled from components.

We assembled a solar generator (inverter, charge controller, deep-cycle battery, and solar panel) in case the gas-powered generator died.

After the storm, we made sure we would not be caught unprepared again. On our first trip to the store, we purchased:

- White-gas lanterns with hurricane lamps and LED lanterns that burn for days on a set of batteries
- A used ATV (just in case)
- Upgraded our chainsaw to a heavy-duty model
- A manual grain mill and nitrogen-packed hard red winter wheat-berries from which to grind flour.

And we're still at it! As we discover new products that appear useful, we budget for them. I reveal our lack of preparation before the storm so that you might learn from it and become more self-reliant soon. You can begin by putting survival items in your vehicle today.

Ice Storms

While relatively rare, ice storms can stop just about everything. The only ice storm I experienced was in Atlanta almost four decades ago. I was attending a tradeshow for *Soldier of Fortune* magazine and staying in a nearby hotel. Coming out of the exhibition hall, the deathly quiet was the first thing that struck me. The next thing that struck me was the sidewalk, which hurt as I lay there wondering if any bones were broken. Sliding and slipping to the hotel, I saw scores of people in the lobby—stranded locals who ended up sleeping on the floor that night. Public and private transportation wasn't running because no one in Georgia had tire chains.

Having a set of tire chains and a pair of strap-on ice cleats for footwear is prudent. And ice melt also makes life easier. Beware of downed power lines, which can be deadly, as live ones tend to whip about like angry snakes.

Lightning

Lightning can strike without warning and without a cloud in the sky, which is the origin of the saying: A bolt out of the blue. If you hear thunder or see lightning, descend immediately if you're on a mountain or hill and avoid tall objects like trees, metal fences, or water tanks. Insulating yourself from the ground can save your life. Sit on something dry, like a rucksack or hay bale. Hug yourself into a tight ball: Bring knees to chest, feet off ground, and tilt head to knees. Lightning can arc, burning you severely if you leave a gap between your limbs, so make sure to stay tight. If you have nothing to insulate yourself, lie flat on the ground, legs together, and arms at your sides. If you are near your vehicle, enter it, stay in it, and do not touch metal. A vehicle's rubber tires insulate the occupants from lightning.

If you feel your hair standing on ends or feel your skin tingling, you are in danger of being struck. Make sure to immediately lie flat. I was once atop a rock outcropping when a thick iron rod driven into it began to buzz. I broke a record for climbing off the rock. Seldom do people get a warning of a lightning strike. If you are hiking, discard metal objects some distance from your party. If technical rock climbing, use your best judgment but do not discard any metal object you'll need later. If you can take shelter in a shed or house do so. Caves are problematic places to shelter because lighting can arc across the entrance, so stay one body length away from it.

Nature can be serene or savage so maintain situational awareness at all times.

4

EARTHQUAKES, FLOODS, DROUGHTS, FAMINE, AND WILDFIRES

If you have a major disaster involving hundreds of thousands, or in this case millions of people, whether it be a natural disaster or an act of terrorism, the first seventy-two hours are going to be totally chaotic no matter what you plan to do.

—Warren Rudman

We didn't start the fire. It was always burning since the world's been turning.

—Billy Joel

Shannon Ewing

Haiti's 2010 earthquake demonstrates that lax building standards can make a bad quake worse.

Earthquakes

Did the Earth Move?

On January 23, 1556, the deadliest earthquake occurred in Shaanxi, China, killing an estimated 830,000 people. Having experienced a couple of minor earthquakes in California and Alaska, I assure you that they are quite unsettling. Preparing for a major earthquake is fairly straightforward. Create a plan and rehearse it. During an emergency (when stressed), most people do what they have been repeatedly trained to do, which is especially important if you have children. Unless you are living in a tent you'll need to evacuate rapidly or shelter in place.

Given the number of variables—type, age, and condition of construction and your location within it—no intelligent recommendation can be made regarding the better choice. If you're in a vehicle in a city, stop your car safely, turn off the engine, and huddle near the floor. The vehicle roof will protect you from smaller falling objects. If you're in a convertible, park and seek shelter in a park or open area. If you're in a high-rise office building, do not use an elevator—if the stairwell is choked with panicking people, crouch next to a desk or strong table. If outdoors, lie flat; you might injure yourself by running. Severity and duration of a temblor play a role in determining your response. It's safest to assume that any quake you are experiencing is the worst scenario. Animals often sense that an earthquake is coming, so if your pets or livestock act tense or anxious, the cause may be an impending quake.

How to Prepare

Designate who is responsible for the safety of the children and elderly and establish a rally point where you can meet to determine if anyone is missing. Because cell phones and landlines probably won't be working, leave handwritten messages at the rally point or another designated location. Your plan may change depending on which family members are home when a quake hits. Rehearse nighttime evacuations, because they require going from unconsciousness to thinking straight and moving as fast as possible.

Store plenty of water (in containers you can carry), because chances are excellent that you won't be getting any from the tap. Store food that requires little or no preparation, as you most likely won't have gas or electricity to cook with. Store a supply of food and clothing (for all seasons) away from your house or apartment building so that it won't be buried under rubble. Keep heavy objects off top shelves and make sure to place top-heavy items—grandfather clocks, statues, or sculpture—where they won't topple and crush someone during a quake. Warn family members to stay away from oversized windows, which can become giant guillotines when the ground shakes. Fireplaces and chimneys also are dangerous.

Almost nothing can be salvaged from a bad earthquake.

Safe Areas during Earthquakes

Explain to your family that corners and doorframes are safer than most other places, and that taking shelter next to a desk or strong table is safer than sheltering under it. Learn how to shut off natural gas and propane and main electrical breakers. Team with another family or two so that you can help each other immediately after a quake, which is critical if someone is trapped under rubble.

After a Quake

After a quake your health may be endangered by broken sewage lines, which can cause a variety of diseases. This may contaminate your drinking water, so filter, boil, or treat it with chlorine bleach or iodine.

Preventing an earthquake is impossible, but you can choose to live in non-earthquake-prone areas. As you might imagine, Alaska, California, and Canada's British Columbia are the most active for earthquakes. The U.S. Geological Survey provides updated online earthquake information. Look at maps of earthquake zones that show major fault lines. Go online to investigate recent earthquake activity and visit: earthquake.usgs.gov/earthquakes/states/top_states.php

Some structures stand while others fall.

Masonry as Shelter

Stone and brick homes are great for surviving fires and protecting residents from small-arms fires (not brick veneer), but even when reinforced with steel, they perform poorly during earthquakes. Research demonstrates that single-story frame houses that are roughly square maintain their structural integrity best during a quake.

Even the maternity hospital's reinforced concrete cannot withstand nature's powerful forces. The flags are of those countries rendering aid immediately after the Haiti earthquake.

Floods

Slow floods enable you to remove important documents and photos from your home or business. Flash floods are particularly dangerous because rain may not be falling where you are, catching you off-guard. Worst case, you are hiking in a narrow canyon and are overtaken by a flash flood. If so, begin looking for roots or trees that you can grab to pull yourself out. Having fallen in a nonflooding creek after slipping, I advise gripping tighter than you think necessary or the rushing water will force you off. If your pack makes swimming difficult, dump it, but keep your shoes on because you'll need them to hike out.

Preparing for floods is very straightforward. Learning the fastest route—by vehicle and on foot—from your house to high ground is essential. When building a structure, don't build in a flood-prone area. Seal basement walls on the outside to prevent leakage. Elevate water heaters, furnaces, and electrical panels. Build an earthen berm on the sides of the house facing a potential flood source. For each family member, make an emergency kit, including bottled water in a waterproof bag (leave air in the bag so it will float) plus a flotation device. Waterproof bags with transparent panels work great because they reduce fumbling.

Preparing in Flood-Prone Areas

Get a rowboat with oarlocks—a used one is fine—and keep it where rising floodwater won't force it upward into rafters, rendering it useless. Inflatable rafts can lose air pressure over time and be cut by sharp objects in floodwater. A boat is essential unless you want to swim or float in floodwaters containing snakes and other critters. Avoid round-bottomed boats because they're less stable than flat-bottomed ones—you'll have enough to do without having to stabilize a canoe or similar craft each time that people get in and out. If children or elderly are in the household, the boat must be light enough so that they can handle it: Aluminum and composites usually are lighter than wood.

Don't pick up stranded or swimming flood victims, as they can swamp your boat. You bought the boat to protect *your* family, but rather than abandoning these victims, throw them a floatation device. When not on the water, don't use the boat as a storage bin: It's for life-threatening emergencies. If you use the boat for recreation as well as flood-caused evacuations, make sure to replace the emergency kits after every use. If the boat sustains damage, repair it immediately, because your family's lives depend on the boat being serviceable at all times.

Don't Drive and Drown

About 400 people in North America die each year from being submerged in a vehicle. If a flood occurs while you're driving, you will expectedly be scared. Who wouldn't? Panic is your enemy in these situations, because it burns through energy and uses precious air, so stay in control. Tell yourself that you can panic—and use your cell phone—after you have reached land. Follow these steps:

- Brace against the top of the steering wheel—the safest position if the airbag deploys.
- Undo your seatbelt while reassuring passengers—you have about a half minute.
- Undo the children's (or elderly) seatbelts, starting with the oldest child first.
- Open the windows the instant the vehicle's in water.
- Break the window if it won't go down—you can buy a punch or hammer made for this purpose.
- Escape through the open or broken window—water will be filling vehicle—use only arms to swim (so you don't kick those following you).
- If the vehicle fills with water (equalizing water pressure), try the door or exit via window.

Drought

In the 1930s, the drought-caused Dust Bowl (mostly in 1934 and 1936) was the source of extensive damage to 100 million acres in North America: Many farmers migrated to more clement climates like California because of this. Drought creates more fuel for wildfires while reducing water with which to fight them. It can lower water levels to the point that fire hydrants run dry. Both 2011 and 2012 were bad years for drought and wildfires, particularly in the west. The Mississippi River dipped to the lowest level since 1988, stranding many boats and barges: The estimated cost of damage was $1 billion. If drought persists over several years, some animal species will migrate to regions having more available food and water. Drought also shrinks crop yields, raising food prices in proportion to the severity of the drought.

Preparing for Drought

Aside from storing water for your family, pets, and livestock, you can prepare for droughts by buying crops and plants known for their drought resistance. Catching and storing rainwater—not legal in some states and counties, don't ask me why—is prudent.

Wildfires

Droughts create the fuel that wildfires need to become major wildfires like the ones that hit the west in 2012. My experience in battling a couple of small-scale wildfires in the foothills of Colorado as a volunteer firefighter taught that prevention beats reaction. Creating defensible space around your property can take hours or years depending on your location. Living in an evergreen forest, I know the ongoing work of creating defensible space: Felling and trimming trees, hauling away slash, and raking up hundreds of contractor bags full of pinecones and needles. Leaving the trees untended is not an option, even if a fire hydrant is close by. In no case should dead trees or shrubs (fire fuel) remain for more than a few days. How long would you leave open tubs of gasoline on your property?

If you are camping or working in the wild and see a grassfire coming toward you from some distance, grab your pack and begin walking away from it. As you walk, look for patches of exposed rock, old foundations, or a body of water where you can wait safely for the flame-front to pass you by. If you don't see a safe spot, begin moving left or right: Grassfires usually run with the wind, so by moving left or right, you may outflank the fire. If the grassfire is gaining on you, dump your pack and pick up your pace. If you're wearing synthetic pants, take them off before the flame-front reaches you. During a fire, synthetics—with the exception of Nomex®—can melt onto your legs, which makes treating and healing your burns more difficult.

If you are near your vehicle when a full-on wildfire moves toward you, grab your keys and run immediately toward the vehicle. If you're on foot and near a good-sized body of water, run toward it. Remember that you're running for your life. Wildfires tend to burn uphill, so run downhill if you can, but know that strong ground winds can push flames downhill, so make sure to pay attention to wind direction.

Structure and Roof Considerations for Fire Resistance

Brick, stone, and stucco structures having defensible space often can withstand wildfires. But understand that burning trees or structures some distance from a house can create sufficient radiant heat to set fire to curtains and other flammable items inside the house. The distance that radiant heat can travel to start fires depends on the number, size, and type of burning tree. For example, juniper trees burn much hotter than most other types of trees because of the resinous sap they contain.

Frame houses often catch fire when adjacent shrubs or trees begin burning, so clearing shrubs and trees away from a house makes a great deal of sense. Some rural fire departments are trained to pass by houses without defensible space so that they can protect those that can be saved.

Steel roofs, slate, fake slate, concrete tiles, and clay tiles can prevent wildfires from getting a toehold on your house. Asphalt (composite) shingles are more fire resistant than shake (wood) shingles. When considering composite shingles, Class A (fiberglass) shingles provide the highest fire resistance, and organic shingles, Class C provide the lowest. Shake shingled roofs offer practically no fire resistance.

Famine

Fires can be fast and furious while famine is usually an event that takes time to worsen. Widespread food scarcity is practically unknown in

North America. Soaring obesity statistics prove most people aren't missing too many meals. Whether due to overpopulation, crop failure, or government policy, famine kills millions worldwide, particularly in sub-Saharan Africa. Famine, like drought, can cause mass migration. The Great Famine in Ireland between 1845 and 1852 caused the death of one million people and saw one million migrate to other countries. Other scenarios can cause famine in North America; which you can read about in Chapter Nine on nuclear and radiological warfare and Chapter Ten about electromagnetic pulse.

Preparation against Famine

Short-term preparations for famine are probably what you'd expect: Store food that your family enjoys and eats regularly to come through six months of upheaval well nourished. Buy extra food for neighbors, friends, and relatives; most of whom will not be prepared for famine or any disaster. Depending on your family's level of physical activity, figure each person will consume 2,000 to 3,500 calories per day—cold winters require more fats in the diet depending on hours spent and work output outdoors. For example, working an office job versus cutting firewood for five or six hours in twenty-degree weather will require different diets. While working north of the Arctic Circle in Prudhoe Bay for a year, my caloric intake nearly doubled, just like in normal times your budget will determine the quality of food you eat.

Long-term planning includes storing sufficient seeds for planting so you'll have fresh food going forward. Store food and seeds in a cool place or at no higher than room temperature to make sure that they'll still be usable when the time comes. Warmer storage temperatures shorten shelf life considerably. Shelf life won't be a problem if rodents or other pests eat your non-canned food, so check on your investment regularly. No new-fangled technology will assist you in surviving a famine. Store food out of sight as you would any valuables.

If the power fails in warm weather, eat perishables (fresh and frozen) first. If the power goes out in the winter, place your food in a critter-proof container and put it in the snow and shade. Consider including convenience foods like canned cheese, canned butter, canned bread, and peanut butter for long-term storage and for keeping in a vehicle emergency kit during cold weather—liquids will freeze. Remove these canned goods each spring or the hotter weather will degrade the nutritional value and severely shorten their shelf life.

Food: A Question of Priorities

When clients hire me to consult about their preparations, the subject of food storage arises early on. If their food storage is inadequate, they often say it's all that they can afford. A brief look around sometimes reveals two or three late-model imported vehicles, a home theater, large flat-screen televisions, original sculpture and oil paintings, a gym, a boat, a hot tub, and snow machines for each family member. I point out their luxurious lifestyle while telling them what they probably don't want to hear: They had the money for increased long-term food storage, but they chose to buy other things with it. Some of my clients swallow hard, sell a few luxuries, and put the proceeds into food and other practical tangibles. Most don't.

Other clients go overboard on weapons and ammunition. A family of four may possess a score of expensive guns and high-end optics as well as tens of thousands of rounds of quality ammo. But their food pantry would not last more than two months. When asked how many weapons they can shoot at the same time, most clients see the light. Achieving the optimum balance of resources is critical for beginners and experts alike. Thoughtful planning costs you nothing but yields significant benefits.

5

ECONOMIC DISASTER: DEFLATION, INFLATION, AND HYPERINFLATION

It is well enough that people of the nation do not understand our banking and monetary system, for if they did, I believe there would be a revolution before tomorrow morning.

—Henry Ford

It is incumbent on every generation to pay its own debts as it goes. A principle which if acted on would save one-half the wars of the world.

—Thomas Jefferson

Money is like blood . . . when you need it, it's critical. We all grew up using currency (money), so we often take it for granted. Money seems to be straightforward: We earn it, save some, and spend or invest the rest . . . and pay the mandatory tax donation to the U.S. Treasury. If you think that's all there is to currency, then please read on. A brief review of facts will illustrate how your money has been tampered with.

The continuing economic decline and eventual failure of our currency is inevitable when all the facts are laid out clearly. In full disclosure, I don't have a graduate degree in economics, finance, accounting, or banking. But critical thinking and experiencing several recessions and recoveries qualify me to discuss monetary policy and the economy. It doesn't take a genius to discover that the United States is the biggest debtor in the world and is bankrupt. Given the following, you can anticipate and prepare for the coming economic crash.

Monetary Magic

U.S. dollars were originally silver and gold coins. Many of these coins remain and have appreciated in value as the price of these metals has risen dramatically during the last decade. Another way to view the price of precious metals is that the metals have maintained their value while the paper dollar's value has decreased. When mentioning the value of gold and silver coins, I am referring to their intrinsic worth, not those having collector value.

A Bit of Background

From 1878 on, silver dollars and paper greenbacks were available and interchangeable. In 1913, the U.S. Congress passed the Federal Reserve Act, which established a central bank and gave it the legal authority to issue Federal Reserve Notes (U.S. dollars) as legal tender. Although it's called the Fed, *it is a private organization* that works closely with the U.S. Treasury Department.

In April and August of 1933, President Franklin D. Roosevelt signed Executive Orders creating the Confiscation Act of 1933 (E. O. 6102 and E. O. 6260 respectively), which made it illegal for citizens to possess gold coins or bullion. Citizens had a few weeks to exchange their gold for Federal Reserve notes at a specified price for gold. If citizens did not abide by the Confiscation Act of 1933, then they were subject to a $10,000 fine and/or a ten-year prison sentence.

In March of 1964, Secretary of the Treasury C. Douglas Dillon ceased the redemption of paper dollars ("silver certificates") for silver dollars—after nearly a century of Americans exchanging silver dollars for paper dollars and back again. Going forward, the U.S. dollar, not backed by silver, was fiat money. Issuing currency by fiat means it has no intrinsic value. It represents government debt.

In August of 1971, President Richard Nixon announced that going forward, gold would no longer back the U.S. dollar—although U.S. gold coins were an established part of the American culture. Other nations also began removing gold as backing for currencies. Money "floated" in relation to other currencies. In time, only central banks stored gold.

Transforming the U.S. Dollar from Real to Unreal

So the U.S. dollar began as currency having genuine value because it was made from precious metal. In a series of Congressional acts and Presidential Executive Orders over several decades, real money

From silver coin to Silver Certificate to Federal Reserve Note, the dollar is no longer tied to anything of real value.

became converted to impressive-looking Federal Reserve Notes having no intrinsic value. What this means to you is that as the national debt increases the dollar's value decreases. "Quantitative easing" increases inflation and raises the prices of commodities (food, precious metals, oil, copper, and steel, for example).

Choking the U.S. Economy

Any of the following elements are compelling and complex enough to write a book about. I briefly summarize them to illustrate the scope of problems given monetary mismanagement and economic decline. Even if half of the following conditions were corrected tomorrow, the unsustainable government debt would prevent the economy from recovering.

Debased Currency

History demonstrates that empires in decline shared a common failing—they debased their currency. As those inside and outside the empire began to notice the currency's diminishing worth, they lost faith in it. Over time, this erosion of faith meant that those outside the empire would offer less and less in exchange for a debased currency. Eventually those within the empire rose up against their rulers or chose to live outside the empire. See the preceding section to discover how the U.S. currency has been debased.

INFLATION: THE SILENT TAX

To understand how inflation affects you, think of a tire on a rim. The tire represents the expanding and contracting economy. As air is added, currency inflates. As air leaks out, currency deflates. Too much air causes the tire to burst (currency hyperinflates) and have no value. The rim represents precious metals, because they are a relatively stable store of value that can accommodate a variety of tires (currencies) over time. Inflation is possible only with fiat currency and has been eroding the U.S. dollar's value for decades. Zimbabwe's dollar has become virtually valueless because of government-caused hyperinflation.[1]

Inflation causes goods and services to cost more, which is an accurate way to track it: You can list products and services that you buy often and record the prices over time. The U.S. government generates statistics about the economy. These statistics would be useful if the formulas for them remained constant, but unfortunately, they don't. The most recent deceit was to change the Consumer Price Index (inflation gauge) formula by removing fuel and food. The government can continue to manipulate the formulas that it uses to calculate economic statistics.

As you may know or suspect, fuel prices have not risen; the U.S. dollar's purchasing power has decreased. Fiat currency maintains its value only if the country's leaders are honest and forthright. When U.S. leaders—the President, Secretary of the Treasury, Chairman of the Federal Reserve, and Congress—irresponsibly spent billions then trillions of dollars, they started the U.S. economy on a dangerous downward spiral. *Worse, they borrowed the money that they spent.* As interest rates rise in the near future, the time will come when the government cannot service the debt (pay the interest on it), let alone repay it.

DECLINING U.S. CREDIT RATING

The dollar is declining. The U.S. credit rating fell from AAA (Best) to AA in August 2011. So if you think the dollar may fail, it follows that investing in stocks or bonds denominated in dollars also could lead to failure. But if you'd bought gold in 2000, your investment would have increased more than 500 percent. Don't take my word for the preceding, but rather do your own research to make an informed decision.

---

[1] In 2009, the Zimbabwean dollar was abandoned. The South African Rand, Botswana Pula, Pound, and U.S. Dollar are now used in its place.

BAD DEBTS TO DEVELOPING NATIONS

America's government has made a number of bad loans (directly and through the World Bank and International Monetary Fund) to developing countries. Many of the debtor nations cannot or will not repay these loans, so they were "rescheduled." Rescheduling is another way of the U.S. saying, "We'll take smaller payments." This means that the taxpayer is responsible for billions of dollars in unpaid loans.[2]

TOO BIG TO FAIL

The stimulus didn't work. While bloated money-center banks like Goldman Sachs received billions from the Bush and Obama administrations, taxpayers derived no benefit. The too-big-to-fail concept flies in the face of free-market capitalism. If an enterprise takes unwise risks or uses poor judgment, it can declare bankruptcy and reorganize. If after reorganizing, it still cannot make a profit, investors can buy the company's remaining assets and begin anew. The marketplace, not the government, has been and remains the best judge of which businesses should succeed or fail. Bailing out financial institutions because they are huge and well connected with politicians is a bad idea by any standard.

SO-CALLED *ENTITLEMENT PROGRAMS*

Entitlement programs began as a true safety net and were well-intentioned. As additional programs and layers of bureaucracy were added over the years, the system became unwieldy. This complexity opened the door to waste, fraud, and abuse. The result is that most of these programs are now bankrupt. Social Security, SSI, Medicare, the Medicare Prescription Drug Improvement and Modernization Act, and food stamps all contribute to America's unsustainable debt.

---

[2]   William Easterly, who is a professor of economics at NYU, stated in 2007 that ". . . We've already spent, as official donors, $600 billion in aid to Africa over the past 45 years, and after all that, children are still not getting the 12-cent medicines (to fight malaria). . . . $600 billion in aid to Africa over the past 45 years, and over that time period there's basically been zero rise in living standards."

PUBLIC EMPLOYEE UNIONS AND PENSIONS

Unions began as a way to protect workers from unsafe working conditions and substandard pay. Collective bargaining enabled disenfranchised workers to improve their lot. When municipal, state, and federal government employees were unionized, however, an unhealthy relationship began between the unions' powerful lobbyists and perpetually campaigning politicians. The unions donated to politicians who were elected and wrote legislation favoring the unions, and the cycle of corruption has been going on for decades. While most taxpayers appreciate and respect police and firefighters, they are disgusted by union contracts and pension plans enabling workers to retire while in their fifties at almost full salary. Worse, many civil servants make higher salaries and have better benefits than those in the private sector.

PORK BARREL PROJECTS

Decades ago, shortsighted politicians learned that spending taxpayer money within their home jurisdiction was a sure-fire path to reelection. But building bridges to nowhere, ego monuments, and cheese museums were not good for the country as a whole.[3] Earmarks, another way of referring to pork barrel projects, continue even as the ship of state is sinking. The taxpayer has paid for thousands of projects that do little for anyone. As Ronald Reagan said, "Government is not the solution to our problem. Government is the problem."

THE U.S. TAX CODE

As of 2012, the U.S. tax code exceeded 70,000 pages, many of which contain vague language, making some rules and regulations open to

---

[3] In 2010, the Triboro Bridge, which connects Queens, Manhattan, and The Bronx, was renamed the RFK Bridge. This cost the city $4 million, and they have since announced the renaming of several other bridges and tunnels.

interpretation. The code also contains outlandish tax deductions for individuals, like buying alpacas. In some instances the government's Overseas Private Investment Corporation (opic.gov) pays corporations to move overseas. At minimum, these vague rules, odd tax deductions, and counterproductive government programs detract from the Tax Code's credibility and fairness while adding to the taxpayer's burden.

U.S. GOVERNMENT SUBSIDIES

Subsidies began when most farms and ranches were family operations; not the agribusiness it is currently. Subsidies are a way of incentivizing farmers and ranchers to grow or not grow certain crops and raise or not raise certain livestock. Like so many well-meaning government programs, unintended consequences began to outweigh the benefits as subsidies grew faster than the crops they were meant to discourage. When farmers discovered they could receive government money for not planting, fertilizing, and harvesting a certain crop, they took the government up on its offer. When dairy farmers found out the government would subsidize milk, they accepted the funding. Subsidies have done nothing less than distort the market. Taxpayers foot the bill for what has become runaway subsidization.

OVERREGULATION

Federal regulatory bodies have expanded their spheres of influence, writing more regulations year after year. The result is that the Environmental Protection Agency, Food and Drug Administration, Department of Transportation, and Occupational Safety and Health Administration are making it more difficult to do business and adding significantly to the cost of all goods and services. This costly overregulation puts the United States at a disadvantage to competitors having less regulated economies.

ILLEGAL IMMIGRATION

Illegal immigration is costing taxpayers billions of dollars each year. Social Security and Medicare withholding taxes unclaimed by illegal immigrants do not begin to offset emergency room visits, schooling, and other government benefits they receive. We have a porous border and the administration has instructed the U.S. Border Patrol not to enforce many immigration laws. This situation must be turned around to reduce the hidden costs of illegal immigration. Make no mistake the problem is unenforced United States immigration laws not the immigrants, most of whom are pursuing a better standard of living for their families.

INCREASING BANK CONSOLIDATION

Since January of 2008, more than 400 banks have failed. The FDIC has stepped in to ensure a smooth transition of a failed bank's assets to a stable financial institution. These bank failures have led to a further consolidation of small banks, most of which received little if any funding from the Troubled Asset Relief Program. The too-big-to-fail banks, bulging with stimulus money, continue to gobble up smaller banks—a disturbing trend because community banks tend to keep money in the community. The too-big banks are deeply in debt.

POLITICAL FAVORITISM

Corporate clout can be enhanced by contributing millions of dollars to political campaigns. When the candidate wins, there is a tendency to give contributors favorable treatment. This favorable treatment is unethical. In some instances an administration provides millions of dollars in grants to certain companies. Some of these companies later declare bankruptcy so again taxpayers pay for politicians' poor judgment. Bankrupt Solyndra is a prime example of political favoritism.

DERIVATIVES: INTENTIONAL COMPLEXITY TO OUTWIT REGULATORS

When the U.S. economy experienced a deflationary collapse in the late 1920s—The Great Depression—businesses were able to climb out of the hole because the economy was smaller and simpler. A depression can be described as too few dollars chasing goods and services. Could the U.S. economy experience another depression? I think the United States is in a depression now, but that the government and media are keeping the public in the dark regarding the real state of the economy.

A few years ago, Goldman Sachs developed derivative instruments, which is a contract between two parties that specifies conditions (especially dates, resulting values of the underlying variables, and notional amounts) under which payments are to be made between the parties. The complexity of derivatives and collateralized debt obligations is so overwhelming that even financial experts and Securities and Exchange Commission regulators have difficulty understanding them. Tens of trillions of at-risk dollars invested in derivatives are casting a shadow over the world's stock markets. One of the few things that financial experts agree on is that if the derivative market becomes unstable, the world economy will grind to a halt soon afterward.

THE COST OF COUNTERTERRORISM

Terrorism remains a threat although the United States has not sustained a major terrorist attack since September 11, 2001. The radical Islamists' choice of targets was symbolically and economically devastating. In reaction to September 11, the Bush administration created the Department of Homeland Security: The 2012 budget was $39.6 billion. The Transportation Security Administration was created after September 11, when it was determined that the patchwork of private security firms providing airport security was inadequate. Founded by Executive Order in 2004, the National

Counterterrorism Center combines representatives from numerous intelligence and law enforcement agencies to facilitate interagency communication and sharing of information. Budgets for U.S. intelligence agencies have doubled since September 11. The Patriot Act is doubly dangerous because taxpayers pay for a program that diminishes their privacy. The preceding reactions to terrorism, however necessary, come at a tremendous cost to freedom.

The Bottom Line

America is a vast country with an inventive, industrious populace, so it can tolerate one or two of the preceding conditions. Taken together, however, they have derailed the economy. These long-standing issues are complex and intertwined, so beware of anyone offering a simple solution. These problems will take years to solve. Leaders and followers will need discipline to right the damaged monetary and economic situation. Public employees demonstrating violently in Greece are a cautionary tale about what could occur in the United States. Our debt currently exceeds *$16 trillion* (usdebtclock.org). To pay the debt, we need to pay even higher taxes and diminish our lifestyle. Then where will we be? Broke.

Preparation for the Savvy and Self-Reliant

While you probably knew most of the preceding conditions, seeing them together is certainly eye-opening. The next steps are about preparing for the trying times that are coming. Numerous books about the problematic U.S. economy have been written since the economic earthquake in the fall of 2008. If you agree that the sum of government (federal, state, and municipal), corporate, and personal debt is unsustainable, choose a book about the damaged U.S. economy that makes the most powerful case and insist that your family read it so they know what's coming.

DECREASE DEBT

While the government and corporations are in debt, you can begin reducing your family's debt. Pay off credit cards if you have money in the bank—especially if you use credit cards as a convenience, not because you need the credit. You can sell items that you no longer use or need to generate hundreds or thousands of dollars and unclutter your life. Possessions pile up because it's easier to store them than to decide whether to sell, donate, or discard them. The Internet is a great source for determining the current value of something that you want to sell. Visit eBay and Craigslist to view recent prices. Take high-resolution photos to help sell your items. Maximize features, benefits and advantages in your advertisement. Keep the goal in mind: You're selling possessions to use the money to reduce your debt, not to spend it on something else.

STOCK MARKET AND U.S. DOLLAR

The stock market has become increasingly unpredictable. Most of the old rules about investing were rewritten after 2008 and 2009. Fiat currency is representative money, and stock certificates represent a fraction of a publicly held corporation: Neither kind of paper has any intrinsic value so either one, given extreme volatility in the markets, can become worthless.

BUY TANGIBLES

One alternative to owning pieces of paper is possessing tangibles, which tend to maintain their value over time: Light bulbs, batteries, paper products, toothpaste, deodorant, razorblades, soap and shampoo, firewood, tobacco products, ammunition, silver coins, tools, crop seeds and fertilizer, clothing, and motor oil. In the event of a currency collapse, bartering will replace fiat currency for a time, in which case you already will own a variety of useful trade-goods.

Few people will trust the dollar after a currency collapse, so the U.S. Treasury will print different-looking currency, perhaps tied to the value of a precious metal or other commodity.

One of the few things that you can be sure of during the next few years is that you and your family will need to eat. So you can reduce food costs by buying sale items in quantity. When you discover a great sale on a nonperishable food item, instead of buying a little of it, buy as much as you will need for a year. By purchasing cases of nonperishable food you accomplish three things: You benefit from the security of having an ample supply of food. You make fewer trips to the market, saving time. You have food on hand in case of a disaster. Make sure to note the expiration date of canned food so that you can use it while it's still nutritious.

You know best what you'll need during the coming year, so make a list and establish a budget for acquiring the items as soon as practicable—during the last decade most wholesale and retail prices have risen, for example, gasoline and diesel fuel prices. Many economists contend that inflation will continue to increase so time is not on your side. If you own precious metals take physical possession of them—don't keep them in a safe deposit box. If your bank closes for any reason, you won't have access to what's yours.

KEEP CASH CLOSE

Keep cash on hand to take advantage of buying opportunities as the economy declines. Cash will be accepted gladly when credit cards, debit cards, and checks may be met with a frown. Also, if a bank announces a holiday—banks will be closed to the public, including customers—the only cash you'll have is the cash that's not in the bank. While a run on a bank—so many customers demanding cash at the same time that the bank runs out—hasn't happened in years, it would happen if the dollar suddenly drops in value. If you're wise enough to avoid holding U.S. dollars, buy foreign currencies that you like, for example, Swiss francs, Australian dollars, or Norwegian krones. But keep enough U.S. dollars on hand to pay for everyday items if the banks close for a week or two. Regardless of which currency you keep outside of a bank, cash will be king.

Investigate other ways to invest in securities that are not denominated in U.S. dollars. Research mutual funds denominated in Swiss francs, for example. However you choose to invest—for example, tangibles, precious metal mining stocks, or physical precious metals—match your investment strategy with your temperament: You'll sleep more soundly that way.

Cash will rule regardless of country of origin.

6

CIVIL UNREST

When bad men combine, the good must associate; else they will fall one by one, an unpitied sacrifice in a contemptible struggle.

—Edmund Burke, Philosopher

Can't we all just get along?

—Rodney King

Civil unrest sounds less life-threatening than rioting, mob violence, looting, and arson. Well, it's not. Describing civil unrest is almost unnecessary because most people in North America have seen it on the nightly news.

In August of 1965, the nation watched the Watts section of Los Angles, California burn during five days of race-based rioting as 34 people died, more than 1,000 were injured, and more than 3,900 were arrested.

Between July 12 and July 17, 1967, the Newark race riots were a major civil disturbance that occurred in New Jersey. The days of rioting, looting, and destruction left twenty-six dead, and hundreds injured.

On July 23, 1967, the Detroit riot began. It turned into one of the deadliest and most destructive riots in U.S. history, lasting five days. To end the rioting, Governor George Romney ordered the Michigan National Guard into the city, and President Lyndon Johnson sent in the National Guard and U.S. Army troops. The rioting left 43 dead, 467 injured, more than 7,000 arrested, and more than 2,000 buildings destroyed.

The Most Destructive Riot in the U.S. in the 20th Century

In March of 1991, motorist Rodney King led Los Angeles police on a high-speed chase through the streets of L.A. After apprehending King, the four arresting officers brutally beat him. A resident shot a ninety-second video of the beating, which was aired repeatedly on national news programs.

In April of 1992, the four police officers were on trial for the brutal beating of King. They were acquitted of any wrongdoing in the arrest. Hours after the verdict, protest and outrage turned violent in South-Central Los Angeles. During three days of rioting and looting, fifty-five people died, 2,000 were injured, 4,000 fires were lit, and 7,000 were arrested. President George H.W. Bush ordered the military to quell the riots, which were the most destructive in the 20th century. The total cost of the damage was $1 billion.

America is a relatively violent nation, more so than, say, Canada. Violent crime as well as the likelihood of rioting tends to increase in low-income neighborhoods, so make sure to maintain situational awareness when entering one. If you haven't seen videos of rioting, they illustrate how quickly and spontaneously mobs form given the right trigger: If they did not form quickly and spontaneously, the authorities would already be there to prevent or end it. Once a riot has begun, it takes on a life of its own. If you are in the wrong place at the wrong time you may die. Read *Understanding Riots* by David D. Haddock and Daniel D. Polsby for a more scholarly approach (cato. org/pubs/journal/cj14n1-13.html).

Mental Preparation for Surviving an Urban Riot

Surviving large-scale rioting requires serious mental preparation. Realizing that civil unrest can happen without warning is the beginning of mental readiness. For example, rioting in major cities can be triggered by celebration (drunken fans setting fire to police cars) and by outrage (a police officer shooting an unarmed teen). Europe's soccer riot participants are mostly young male fans who go to the game intending to drink and fight. Regardless of the cause, your response is the same: To help your family escape the rioters. By knowing the preceding you will be less surprised if you find yourself in a fast-forming mob. Prevention beats reaction so staying off city streets after the World Series or the Super Bowl is prudent.

Physical Preparation for Surviving an Urban Riot

The following tactics assume that you're fit enough to run—part of surviving any situation is staying in good physical condition. Physical preparation begins with always wearing shoes or boots that you can run in. For women this means wearing flat- or low-heeled shoes with straps (lace-ups are best) or carrying lightweight running shoes at all times. Regardless of footwear, you must be fit enough to run far and fast enough to elude the swiftest person in a mob. The race won't be very long because as the mob thins, participants lose their nerve.

Steel-toed boots can prevent crushed toes during a riot.

Author wears pack in the wilderness and urban areas for the same reason: It leaves his hands and arms free.

Wearing a small backpack or fanny-pack is better and safer than carrying shopping bags or a purse, as it's easier to run, maneuver, and defend your family with free hands. If you have small children with you, have them grab your belt with both hands and hold on. You should be able to drag them to safety if you do not have to stop for them or pick them up. Their best chance for survival is if you stay upright and balanced. Practice this technique at home. Make it fun but make sure the children understand that they must hold on, no matter what.

If the mob is far enough away, climb on top of a tractor-trailer and lie flat on the trailer's roof. While rioters will likely smash the windows of the truck's cab, they won't be able to see you. In the worst case scenario, preparation includes being able to defend yourself and your family against an unarmed or armed assailant. If you cannot, take self-defense training or a hard-style martial arts course like Tae Kwon Do. Better yet, train and apply for a concealed-carry permit and carry a handgun.

Stampedes Are Spontaneous and Deadly

Fleeing a burning building, nightclub, or stadium riot is difficult to prepare for because of the sudden panic. In November of 2010, more than 350 people died in a stampede on a bridge in Cambodia. The YouTube video of the aftermath shows people jammed so tightly together that they could not move. Compressing or confining people increases their panic. In February of 2003, the E2 nightclub stampede in Chicago killed twenty-one people. In February of 2003, a fireworks accident in a nightclub in Asbury Park, New Jersey, killed ninety-five people because the doors were locked, causing crushing deaths by panicking people. In December of 2009, a fireworks accident in a nightclub in Perm, Russia, killed 139 people. My advice is to avoid crowds, especially where exits are few or blocked. Note the location and number of exits whenever entering a public place.

Flash Mobs

An urban flash mob is impossible to predict and difficult to avoid unless you are part of the specific social media network. If you are shopping away from a store exit when a flash mob or flash rob (intent on crime) enters, you have little choice but to stay where you are and avoid eye contact with participants. Most flash mobs form quickly and disperse quickly, so remain calm and vigilant and be ready to defend yourself. If you are near a store exit when a flash mob enters, work your way to the exit and leave. Most flash robbers want nothing more than to get away quickly with whatever they've stolen.

Chemical Deterrents

If it is legal where you live, buy a large (13 oz) can of bear spray, which ranges in price from $32 to $55. Do not use a small (keychain type) can of pepper spray, as it does not contain sufficient spray to stop initial assailants and deter others. Spray the closest assailant in the face and sprint away from the mob. When the first few spray victims double up and begin coughing and moaning, the remaining members of the mob will likely rethink the situation; if not, turn, stop, and repeat; then sprint away again. Given an urban riot, I'd choose a large can of bear spray over a handgun any day. If bear spray or pepper spray is illegal in your city or state, investigate other spray products that have a similar effect but will not permanently injure others.

When Cunning Beats Running

Tactics for the elderly, injured, or disabled can be as effective as escaping on foot. For example, if you see a mob running toward you from a block away, consider climbing into a commercial or industrial trash container and covering yourself with the contents. Another tactic is to carry a folding white cane and pair of dark glasses with you. When you see trouble coming put on the glasses and walk away from the mob—the blind don't run—while sweeping the cane from left to right. If traffic is moving near you ask a driver for a ride—wave cash to entice drivers.

If you have sufficient upper-body strength, consider removing a manhole cover to hide or escape. Make a manhole-cover tool by winding a few feet of tough wire into eight-inch lengths to connect the centers of two four-inch-long bolts to form an I-shape. Select wire and bolts with care because they must be both strong enough to hold a manhole cover's weight and small enough to slip through a hole in the cover. Carry the tool with you when in a city. To use it, push one bolt through a hole in the cover and use the other bolt as a handle to remove the cover by pulling on it. Knowing how easy it is to make this little tool is worth remembering.

Maintain Your Situational Awareness

Is there a fire escape ladder low enough to jump to and escape? In any case avoid liquor stores or retail stores with jewelry or televisions in the window; they'll be the first to be looted. If you see military troops arriving assume that marshal law has been imposed. Can riots or insurrection happen again? Ask people with firsthand experience: Those in Ireland, London, Spain, the former Yugoslavia, or Greece. Ask those in Lebanon, Iran, Egypt, Yemen, Libya, or Syria.

Can Civil Insurrection Happen in the U.S.?

Rioting in America's cities demonstrates that once begun, violence, looting, and arson often continue for days unless military troops are ordered in. The United States has become increasingly polarized politically. Almost half the country considers itself conservative, and nearly one half identifies as progressive. The differences between left and right are for the most part irreconcilable, which explains why rhetoric during campaigns has gone from mostly respectful to increasingly disrespectful. Ongoing Congressional gridlock on social and fiscal issues is frustrating taxpayers. Almost the only area that the House and Senate agree on is spending more money, which further increases voter frustration.

Consider this chilling scenario:
The U.S. economy continues to unravel as the global economy worsens. Unemployment rises, especially among

younger job seekers. Food prices surge as the drought plus high fuel prices pinch consumers. The mood darkens among unemployed youth. Within the same week assassination attempts on the President as well as the Chairman of the Federal Reserve are thwarted, leaving the Chairman wounded. The Federal Reserve announces another round of buying U.S. Treasury bonds, so-called *quantitative easing,* reducing the value of the dollar relative to other currencies. In response, China begins liquidating the trillions of dollars it has amassed over the years and stops buying Treasury bonds. Japan, second-highest holder of U.S. debt, fears getting stuck with a diminishing currency so it dumps dollars in reaction. Out of self-preservation other nations, even staunch allies of the United States, begin dumping dollars and buying more stable currencies. The dollar plummets worldwide. In reaction, the New York and American Stock Exchanges suffer a massive selloff—computers cannot keep pace—trading is halted.

The President, with the Treasury Secretary at his side, holds a hasty press conference. He reassures overseas investors and the American people that the dollar remains sound. He says that because China holds so much U.S. debt, "It's only natural to adjust its currency holdings from time to time." His confidence doesn't ring true. Usually glib and charming, he seems off balance. He's sweating. Reporters from financial media pepper the President with questions. The President turns and leaves the podium as reporters continue shouting questions. Television programming is preempted. Talking heads speculate about the greenback's future and what other countries will do next.

Customers begin lining up to withdraw money from banks and brokerages. Bank computer networks crash due to the sudden overload of customers banking online. Lines of concerned customers spill into the street. Rumors spread that banks are limiting withdrawals to $200. Television news programs show security guards herding shouting bank customers outside as the ornate brass doors of a downtown bank close. A brick breaks a Citibank window in New York—a firebomb ignites a Wells Fargo bank in Los Angeles. Gunfire shatters several windows of a Bank of America branch in Detroit. As the riots expand to Chicago and Houston and Atlanta, mayors and governors request assistance from the President. Hours go by as the President and cabinet dither.

That evening National Guard units from seven states plus three active-duty U.S. Army units are dispatched to more than a dozen cities where arson and looting are increasing. In Detroit, a soldier is shot dead as he appeals for calm using a bullhorn. His fellow soldiers return fire at the muzzle flash. A raw recruit shoots a pregnant woman in the head. The crowd erupts, overwhelming and disarming the soldiers. Television crews transmit the escalating bloodshed to the networks, which broadcast unedited live feeds in real time. Viewers stare in disbelief as law and order degenerate into anarchy. Firefights are breaking out in most major cities. The Second American Revolution has begun as most of the nation huddles behind locked doors in darkened homes. Unlike the original American Revolution, this one is accelerated by communists, Islamists, anarchists, and extremists who have been waiting years to "even the score," whatever that score is. The speed with which the preceding scenario unfolds leaves little time for city dwellers to execute a plan—if they have a plan.

Riots and Civil Insurrections While Traveling

Before planning an overseas trip, do sufficient research to make certain that your destination country and neighboring countries are safe. The U.S. State Department has an updated online list of Current Travel Warnings (travel.state.gov/travel/cis_pa_tw/tw/tw_1764.html), and the CIA World Factbook contains useful information (cia.gov/library/publications/the-world-factbook). If a country or region is unstable politically or economically, but you must travel there, listen to advice from locals. For example, if they tell you to walk only on a certain street, don't try to second-guess them. The preceding comes from my experiences in Brazil and Mexico.

Prior to departing for a foreign country, get an up-to-date map—not a GPS device because it's breakable—and keep it with your passport. Your passport must remain with you while traveling. In case you become separated from your map, memorize the direction of neutral and friendly countries so that you know which direction to travel toward. If civil unrest begins, you will at least know which border to head toward. Carry a small compass with you—a large compass is better, but you're less likely to carry it. Carry a couple of large silver coins with you with which to streamline border crossings and conversations with bureaucrats and law enforcement types. If trouble starts, don't wait to begin your escape: It's better (safer) to be too early than too late.

Experience traveling abroad has taught me the benefits and advantages of packing light. If you have to depart unexpectedly because a mob is heading your way, you cannot lug a large suitcase, and you probably won't have time to sort through it to discard unnecessary items. Do your sorting before you leave home. Consider buying a piece of luggage incorporating backpack straps—in a pinch you can don the pack, keeping both arms free for running, balance, and defending your family. Chances are excellent that you'll enjoy an uneventful trip—if not, you're mentally and physically prepared for the populace to rise up while you're on vacation. For advice on which items to pack, see Chapter Eleven.

Author travels with luggage containing pack straps in a zippered compartment for versatility.

7

WILDERNESS

… the rope yanked viciously, cutting the harness into his body, bringing a sea of bright-coloured pain; he was suspended over a black, bottomless chasm . . . he scaled the rope, knot after knot, and, with a wild flailing kick, thrust himself into the snow . . . He fell into a faint and lay unconscious . . . his hands bleeding into the snow.

—Lennard Bickel, *This Accursed Land*

Polar exploration at the turn of the 20th century was unbelievably unforgiving and often deadly. The harsh consequences of mistakes are what make the wilderness such a great teacher. If you're seeking a one-size-fits-all approach to wilderness survival, you won't find it here. In fact, you won't find it anywhere, because the number of variables is incalculable. Where in the world will you be? Arctic, desert, sea, tropic, temperate, lake, swamp, mountain area? How much drinking water, food, or clothing will you have with you? How many medical supplies, weapons, or tools? Will you be alone or have to care for others? Are they injured? Are you injured? Are you eighteen or eighty-one? Will you speak the same language? Are you in a hostile environment? How long must you survive? When dreamers fantasize about surviving in the wild they're in peak health and have their expensive knife and favorite sidearm. They don't fantasize about having to work while suffering from severe diarrhea and dehydration, a broken collarbone, or a spiking fever.

If you don't think that you'll ever need to retreat into the wild, consider the following scenario. A deadly virus breaks out in Hong Kong and quickly spreads to New York and London. You watch televised news as the epidemic becomes a pandemic and hundreds of thousands are dying. To safeguard your family, you load the car and return to a remote campsite you stayed at years ago. Do you have the appropriate gear for the time of year? And the knowledge to camp out and feed your loved ones until the virus runs its course?

It may seem counterintuitive, but a great way to learn camp-craft and survival skills is online at YouTube. Using the printed word to describe dynamic skills is less than ideal. You can learn better and faster by watching a video. If you do not have access to a computer and the Internet, a public library can provide it.

If you're planning a trip into the wild to hunt, fish, camp, or work, you will need to be prepared. For example, you know that ticks carry Lyme disease and Rocky Mountain spotted fever. You know enough to walk around a log instead of stepping over it to avoid the possibility of snakebite. You keep vinegar in your vehicle to use on your dogs in case a skunk sprays them. You plan prudently, carry spares of important items, and anticipate likely contingencies. But problems arise when you're unexpectedly thrust into the wild without equipment—that's when you discover how resourceful you are. For example, you take the family to Guatemala for a vacation. The bus you're riding in breaks down in a wilderness area where no cell phones work. As night falls your children are becoming fearful. What now? If each of your packs contains drinking water, convenience food, insect repellent, extra clothing, and lightweight, metalized plastic tarps (Space Blankets), you're all in fine shape.

Render Help When You Can

If you are or are becoming an accomplished outdoorsman, I ask you to help those in distress. While summiting Colorado's Longs Peak, my climbing partner and I got caught in a thunderstorm. Seeing lightning, we pulled on our rain parkas as we ran from the flat top of the mountain. Descending, we came across a family from the Midwest who had attempted the 14,000-foot mountain without adequate clothing. They all wore cotton tee shirts and shorts—once wet, cotton stays wet, sucking heat from the body. They also had no rain gear and were soaked and shivering as they down-climbed a fairly steep section of rock. The children were shivering and verging on hypothermia: Low body temperature caused by exposure to the cold. Shivering is the body's way of producing heat. We helped them down the rock face by climbing alongside, reassuring and encouraging the children and placing their shaking feet on good footholds to descend safely. We didn't need to explain to the parents that they were ill prepared for attempting a high peak that receives thunderstorms almost daily. They were still shivering and thanking us repeatedly as we continued descending.

Author traversing easy rock after ascending a route on the Wind Tower, Eldorado Canyon, Colorado.

Preparing Your Vehicle for Emergencies

Keeping a bag of tools, equipment, and supplies in your vehicle is just plain prudent. The farther you live or travel from civilization, the more comprehensive your bag must be. No one plans to be stranded, but those who anticipate an emergency are far ahead of those who think it won't happen.

Water, Clothes, and Food

Water probably is the first item on your list. You can drink it, wash an injury, or fill your radiator or windshield fluid reservoir with it. But freezing temperatures will turn your supply into a useless lump. If you are indoors part of the day, the solution to having usable water is to carry it with you. If you're outside all day, you'll have to include a stove and pot (compressed gas is less fussy than those using liquid fuel) in your bag or pack so you can melt snow and ice.

Extra clothing is important for a couple of reasons:

1. You'll have something to change into after changing a flat in the mud.
2. It will keep you warm if your vehicle betrays you. Wool is a great choice because it keeps you warm even if it gets wet. Merino wool gets my vote, but it's more expensive. Synthetics dry more quickly than wool, and many who work outdoors prefer them.

Food is more important in winter because burning calories keeps you warm. Convenience foods trump those that need heating—your stove can break or run out of fuel—you needn't wait to eat food bars, canned cheese or peanut butter. Fats are more important during extreme cold weather.

Communication

Communication is important and becomes frustratingly more problematic the farther you stray from urban infrastructure. There's nothing like communication with others to reassure you and turn a life-threatening situation into an inconvenience. The most realistic way to plan is to assume your phone or radio won't work: Such planning prevents disappointment and feeling helpless. Seek a communication device that will operate in the most remote areas that you frequent.

Stay or Walk?

Staying with your vehicle is wise because rescuers are more likely to see a vehicle than a person, and you have shelter from the elements. But sometimes it's wiser to snowshoe or hoof it. For example, if you slide off a backcountry road under dense evergreen trees, your vehicle may not be seen or found for several days. Tire chains, a winch, or traction devices (cat litter) may get you out of mud or snow. If not, leave a note inside your vehicle that includes name, contact information, date and time, plus intended direction of travel and destination—use a pencil or permanent marker because many types of ink run and smear when they get wet. If you are with others, list their names and contact info as well.

Walking out involves risk but is an option if you:

- Are uninjured, fit, used to being outdoors, and adequately clothed
- Know the area or distance to a residence or business
- Have sufficient equipment to reach your destination safely
- Remember a recent, local weather forecast for fair weather.

Why Store Survival Items in a Backpack?

Keeping emergency items in a comfortable backpack is a great decision, because it frees your arms and hands if you decide to walk out to safety. Be selective about which items you carry, as unnecessary weight will slow you. Road flares are worth their weight because they can be used as signal devices and seen at considerable distances. If you're unarmed, flares can also be used to fend off animals. Worst case, you'll need a sleeping bag, pad, bivouac sack or tent, and LED flashlight or headlamp; many of the items used when winter camping.

Other items for a vehicle are up to you because you best know your territory, climate, and limitations. Being a careful sort, I keep two truck boxes on my flatbed pickup. They contain a lightweight axe, backpack, bow saw, cable come along, camp stove, first aid kit, gloves, jumper cables, road flares, four-way lug wrench, paper towels, propane torch, hand and power tools, oil, two-inch ratchet straps, sleeping bag, spare batteries, tarp, nylon tow rope, trailer hitches, and miscellaneous spare parts and fasteners. Plus motor oil, radiator coolant, and a can of flat fix. A jack, lug wrench, spade, fire extinguisher, and fuel container belong in every vehicle.

You also need a cell phone charger that works in your vehicle. Maps and a compass are useful if the shortest way to civilization is through backcountry. Broken-in hiking boots beat new ones. A large can of bear spray is an excellent deterrent for all predators. A handgun is reassuring if the state you're in permits it. A holster can be problematic if your pack has a waist-belt so wear your pack when trying on a holster. A radio provides weather updates, which can alert you to storms. Carry as many tools as is practical because you may have to clear a downed tree blocking your way.

Two steel boxes on author's flatbed truck contain tools and emergency equipment. Yellow containers are for diesel fuel.

A four-wheel-drive vehicle is hands-down better than a two-wheel-drive vehicle. Survival gear includes your fuel tank, which should be filled when the gauge hits the halfway point. The preceding list is intended to stimulate your curiosity about what else would be useful if your vehicle quits in the worst imaginable place. Only your creativity limits the list of practical items to carry. And in no case will products substitute for critical thinking or mental toughness. Anticipate your family's needs in case your vehicle suddenly goes silent.

Natural Shelter

Shelter is essential because it protects you from the elements—as importantly, it's a refuge where you can think and plan and heal if injured. Your physical condition, the weather, temperature range, elevation, and terrain will determine where you shelter. Unless you carry a sleeping bag, pad, and bivouac sack or tent, your need for shelter is paramount. The best shelter is a natural one because you needn't waste time on it. For example, it's easier to spend the night under a large evergreen tree than expend calories building something. And it's more efficient to sleep under an overhanging rock than construct a lean-to. Caves can be great depending on those you share it with. Hornets, snakes, bears, skunks, and mountain lions make rotten roommates. Seek natural windbreaks like hills. Dead branches are great for building but lousy if they fall on you so beware of hinged or hanging branches and standing-dead trees. Avoid ant-hills and other animal entrances and exits.

Built Shelter

Building a shelter depends on the region and resources—incorporate natural elements like trees, logs, and rocks to conserve energy. At a minimum you need a windbreak and roof, and colder weather necessitates insulation. A tarp drastically cuts the time and effort needed to make and break camp. Whatever you decide to construct, make sure the building materials are close by—carrying or dragging them wastes time and energy, which you may not have much of. Nylon cord or vines are necessary to join branches, which may be why parachute cord (paracord) bracelets and key fobs are selling so well.

Shelter and Camp Near Running Water

Shelter near running water when practical but camp high enough so that upstream rain doesn't flood you. Also, make sure to camp about ninety paces from a creek or river to avoid insects and morning dew—never camp in a dry riverbed. Sunlight is desirable unless you're in the desert in summer. Figure out where you can have a fire safely, which warms you and discourages most predators.

When you need overnight lodging, build a debris hut if the debris is not already home to insects or spiders. Understand that debris will not withstand high winds and is basically a tinderbox waiting to catch fire. The steps to build the hut include:

1. Mark a perimeter on the ground just big enough to hold your pack and you when reclining.
2. Dig a 10-inch-deep trench, in cold weather dig deeper.
3. Line the trench with dry branches and leaves.
4. Cut branches to make a framework.
5. Cover in 2 to 3 feet of litter: leaves, branches, and twigs to help retain your body heat.

A better way to learn how to build it is to enter "debris hut" in YouTube and see a hut being built. While that will help, the best way is to grab a hatchet or machete and go out and build one yourself. Lean-to shelters can be built by lashing together whatever you can find in your locale.

A Hasty Hammock

A hammock is fast to rig and beats sleeping on cold ground where your body loses more heat through conduction than convection.

A hammock also keeps you away from snakes and small nocturnal critters. If you don't have time to build a shelter, make a hammock using stout nylon cord and a strong tarp. If the tarp has grommets, loop or tie the cord through them and tie the other ends to a tree. If the tarp is without grommets, knot the ends and loop or tie them with cords and secure the other ends to a tree.

In Cold Winter or Arctic Climates

In cold winter climates, include food that won't freeze and requires little or no preparation—you'll have enough to do without fiddling with a temperamental camp-stove when your kids want food this minute. Snowshoes are not just for recreation so include them: Walking through waist-deep snow is exhausting. You'll need a stove that is robust and reliable—if you don't want to fiddle with liquid-fuel stoves, propane canisters are a good alternative. Carry spare v-belts or serpentine belts for your engine, and change the belts on your engine every year.

Don't exert yourself to the point of sweating, which can chill you and leave you vulnerable to hypothermia; practice limiting your activity so that you don't sweat. In time it can become a habit. If you don't have sunglasses, cut slits in a bandanna and tie it around your eyes to prevent snow-blindness.

Enjoying the snow in Colorado. Note tip of red snowshoe protruding from snow.

Making Snow Shelters

Making a snow cave is a great skill to teach children, as most of them build snow forts anyway. But something has gone drastically wrong if you have to make one to stay alive. Using a wood- or snow-saw is much more efficient than using a shovel to build a cave, but few people carry a saw in their pack. YouTube videos can show you how to build one better than my describing how to do it. So sit down with the family and watch the videos, many of which are brief and worthwhile. Even painfully amateurish videos can contain useful information. Burn a DVD copy of YouTube videos that you find particularly useful. Igloo (Eskimo for house) building takes expertise best viewed on a video so go online to YouTube. If someone near you teaches a course on how to build one, take it.

In Hot Summer or Desert

In hot weather, water is critical. A consulting client told me about driving from Colorado to California without extra water. A quick look at a map shows the folly of driving across large deserts without gallons and gallons of water. If you're stranded, hats with brims, pants, and long-sleeve shirts will keep you and your family from getting sunburned. A bandanna can be wetted to cool those who are overheated, sunglasses will reduce eyestrain, and mosquito repellent will help prevent a sleepless night. If you find or can make shade, shelter there while the sun is shining and travel at night. Avoid eating protein, which uses metabolic water—carbohydrates are best when water is scarce. Scorpions and spiders can be avoided by shaking out your footwear before putting it on.

Desert nights are surprisingly cold, so if you are not traveling at night, seek shelter while there's daylight. A rock wall with an overhang is ideal for a shelter. Gather loose rocks to enclose the overhang. A rock and sand wall can be hollowed out to make a hasty shelter. Digging a pit or trench shelter is possible, but consumes many calories and requires sticks and a poncho or tarp to make the top. You will need a stick or shovel to dig with. Use a cooking pot or pan to dig with if it's all you have.

Rainforest and Jungle

Rainforests are mature ecosystems with high tree canopies, leaving the floor mostly unobstructed. If the mature trees are felled or die, a jungle of shrubs, vines, and smaller trees develops. Rainforests and jungle environments vary widely according to vegetation and animal life. If you plan to be in a tropical environment, double your research, because rainforests and jungles contain plants and animals that can be deadly. The list of threats includes: Feline predators, venomous and constricting snakes, venomous spiders and insects, crocodiles, and alligators. Having lived in Asia and Australia, I have gained a healthy respect for venomous vipers.

While primitive bush-craft is great to know, no prudent person intends to use it in the wild. Finding a forked branch from which to make a fish spear is a great skill, but carrying a fishing line and net around makes more sense. Knowing how to weave leaves to make a water-shedding roof is great, but carrying a plastic tarp is more efficient. So if you want to learn how to make a bow and arrow or a spear or start a fire using primitive ways, go online or better yet, find a wilderness instructor who teaches primitive skills. Some television programs show survival experts using primitive skills, living off the land knowing that they'll have to improvise a shelter and hunt for something to eat. Honest survival instructors will admit they perform on camera to make a name for themselves.

Abundant rainfall is what creates jungles or rainforests, so finding dry tinder and firewood is difficult. While water abounds, it is not drinking water, which means you'll have to boil, filter, or treat it chemically with chlorine or iodine. If you can, use a poncho or tarp to funnel rainwater into a container. Hey, it beats drinking iodine-flavored water.

Night falls quickly in rainforest and jungle so prepare your camp well before sunset. A hammock keeps you off the jungle floor where reptiles slither and critters crawl. Survival without a machete is nearly impossible, and without a net or mosquito coil, you run the risk of getting malaria, which can be deadly. The truth is, you're more likely to be knocked down by a disease-carrying insect than a man-eating predator. Mosquito coils were the defense of choice in Vietnam.

Travel on Inland Waters

Canoes, kayaks, and rafts will get you into and out of the wild when walking cannot. Life jackets must be worn—not used as cushions— to be effective. Waterproofing everything is a must. The good news is that dry bags have become better and cheaper, and ones with transparent panels reduce fumbling for items. While you may consider buying inexpensive food, using inferior dry bags can jeopardize your entire journey.

Knowing cardio-pulmonary resuscitation is mandatory, especially when navigating rapids. Completing a Wilderness First Aid, Wilderness First Responder, or Wilderness Emergency Medical Technician course will significantly increase your crew's survivability if a craft capsizes or a paddler gets bounced out of a raft. Communicating with transceivers is easier and clearer than shouting, especially near whitewater. Scout all rapids because changing conditions can surprise even expert river guides. I was on a canoe trip in western Colorado where the powerful river current wrapped a composite canoe around a rock so that none of the guides could budge it.

When You Are Your Transportation

Whether you are setting out on a planned backpacking expedition or your vehicle breaks down, leaving you stranded, having the right equipment can make all the difference. A vehicle break-down in the wild leads to one of two choices: Stay with the vehicle and wait for help or—if you are low on food and water and know where help can be found—set out on foot to find it. The answer will depend upon your health, age, fitness level, number and condition of passengers, weather, and distance to your destination. If a passenger is too old or young or is not able-bodied, your decision is more complicated.

The following are practical pointers that the wilderness has taught me during nearly fifty years of backpacking and camping. Buy simple, quality equipment that isn't made in a developing nation. If an item fails at home or work you can fix it or buy a replacement: If it fails in the field you may fail as well. A quality used piece of gear beats a new mediocre one. If you get a new item that is crucial for your well-being, test it on an overnight trip—if it fails you know quickly. Multipurpose gizmos usually don't do anything well. Given a choice between a plastic item and a metal one, choose the latter. Each of the following items takes time to evaluate. Don't take my advice or anyone else's without doing your own fact checking.

Boots

If you're wondering which boots are best, the answer may surprise you: It depends. What will you use them for? Where do you live? During which seasons will you wear them? How many years do you want to use them? The final question leads to my bias of buying the best possible, even if you must eat rice and beans for a spell. Many survivors have had to walk out of their predicaments, so it makes sense to wear boots or shoes that you can spend days in without developing blisters or worse. The twenty-six bones in your foot will thank you. Whatever pair you buy, take the time to maintain them with quality products.

Well-worn Raichle hiking boots.

Tromping through Louisiana wetlands, hiking the desert, and backpacking in British Columbia all require different boots, so I'll briefly discuss a few types and their applications. Leather lace-up boots are probably the most common choice in four-season climates. They can be categorized in several ways, including: moccasin toe, plain toe, insulated and noninsulated, plus rebuildable and non-rebuildable. Manufacturers position each type for a particular market segment. The results are so-called: hiking, hunting, logging, tactical, and work (protective-toe) boots. Can you work in a hiking boot or hike in a logging boot? Yes. What really matters is fit and quality. A great fitting pair of inexpensive boots beats an ill-fitting pair of expensive boots. Rebuildable boots have a Norwegian welt, meaning the sole and upper are stitched together on the outside. Non-rebuildable boots have the sole and upper joined so that removing the sole (to replace it) damages the boot.

The National Fire Protection Association approves all-leather White's boots for fighting wildfires.

Since the advent of Vietnam-style jungle boots, textiles (mostly nylon) have replaced leather above the ankle to lighten boots and facilitate drying. Some designs are too complex to be durable—the many seams are prone to ripping and leakage. Gore-Tex® and similar membrane liners can keep feet dry even if the boot leaks.

Pac boots are a hybrid of a leather top stitched to a rubber shoe. Cold-weather models have a felt liner (Sorel® is a well-known example). The boot originated in parts of North America where the terrain is wet for a good part of the year. The downside of pac boots is that the rubber uppers (or shoe) slow the evaporation of sweat, which can cause cold feet in winter and clammy feet during the rest of the year. If you buy pac boots, buy non-felt liners, which tend to self-destruct.

Inexpensive imported pac boots come in lace-up and zipper styles and a variety of heights. They are best for wet areas.

Pull-ons include cowboy, engineer, structural-firefighter, riding, and rubber boots, as well as waders. They are convenient because they're fast to don, but they cannot be adjusted for thin and thick socks. And, in general, do not fit so well as lace-ups. If you need to jump into action in bad weather, they're great. If you have time to lace up, buy lace-ups.

Pull-on boots have their uses but are not so versatile as lace-ups.

Soles come in scores of types for different uses. When looked at objectively, soles are the only part of the boot that provides traction; or put another way, they can make the difference between tumbling down a steep hill and walking down. Vibram® began as a small Swiss firm making high-carbon soles for mountaineering boots. The company has grown to an industry-dominating global giant providing different tread compounds and scores of configurations of high-traction soles to numerous boot manufacturers. Put some thought into your requirements.

Socks

Socks are usually an afterthought. You've invested in a great pair of boots, so why spend money on socks? Thinking that way is a big mistake. For example, it isn't just boots that keep your feet warm during the winter. If you buy your boots a bit large, you can wear a couple pairs of socks, which will insulate your feet and keep them warm. Socks can be bought as anklets, over the ankle, and over the calf. Anklets aren't very versatile, so I don't use them. But the real issue is what socks are knit from: Wool and silk are great natural fibers, and polypropylene is a fine synthetic. I have a few pairs of bamboo socks and find them to be cooler than wool and dryer than cotton. What about cotton? I avoid it because it retains moisture. What about nylon? Nylon tends to refrigerate—that is it gets cold and stays cold—so it isn't what you want surrounding your feet. Nylon is great for reinforcing the heel and toe. Again, buy the best you can. The only things protecting your feet from your boots are your socks. Experiment with cushion-sole socks. Wool and poly pro blends make sense because wool keeps you warm when wet, but dries slowly, while poly pro dries quickly.

Your comfort will increase if you own several pairs of quality socks of various thicknesses and fiber blends to go with different boots. One way to prevent blisters is to wear a thin pair of silk liner socks under a thicker pair, enabling them to slide against each other instead of chafing your feet. What happens when socks wear through?

Darning (sewing) see-through socks is a waste of time. Swallow hard, throw them away, and buy new ones. Or you can wear mended socks and develop hot spots, which eventually become blisters.

A friend of mine, who is a master outdoorsman, has a different approach to socks. He buys one brand and thickness and wears them in the field and on the street. This one-sock method certainly simplifies matters, which is always a good idea. Whichever way you decide to go, treat your feet like they're the only pair you have.

Packs

Choose a backpack that is made of heavy-duty fabric, is comfortable, and uses few zippers, as they cannot be repaired in the wilderness (large safety pins are not an acceptable field repair). To be fair, newer zippers take a lot more abuse than earlier ones. If your pack closes with a drawstring, which breaks, you can replace it with a shoelace. A clean pack design reduces snagging when hiking through dense brush and scrub oak, so choose a pack large enough to contain large loads without your having to lash things to the outside. Internal compartments, while great for organization, limit your options for carrying bulky items. A single open compartment provides optimum load flexibility. Single-compartment packs can be used to cover and warm feet and calves if you have to bivouac. Placing items in color-coded bags enables you to organize and load and unload in moments. Some internal-frame packs use stiffening stays, which can be removed to use as splints for broken bones.

Don't buy a pack for a single feature: Consider air circulation on your back, overall comfort, features, weight, and price. If a pack doesn't circulate air to your back, you'll regret it in summer and winter. Soaking your back in sweat can lead to chilling and hypothermia in winter. If you buy an ultralight pack that isn't comfortable, what have you achieved? If you save a few dollars but regret not having the features you really need, what have you saved? Most retailers allow a prospective customer to fill a pack with sandbags, try it on, and walk around the store with it. Earth tones facilitate concealment while bright colors are easier for rescuers to spot. Talk to experienced backpackers about which packs they like and why. They may mention features, benefits, or advantages that you haven't thought of.

Knives and Cutting Tools

Randall Made Knives® model 19 Bushmaster purchased in 1967. Note smooth, Micarta handle and brass guard. Author drilled hole in handle for a lanyard with cord lock. Sheath features a compartment containing a honing stone.

When buying a knife, hatchet, or saw, consider the use, type of steel, and heat-treating. A smaller knife (4 or 5 inches) will perform as well as a larger one (6 to 9 inches) for most tasks. Fixed-blade knives have no moving parts, so I prefer them. If you use a knife for cutting and splitting kindling, then by all means get a larger one. Avoid so-called survival knives that can be the size of short swords and weigh as much as a hatchet. If working over deep water, a lanyard on

Spyderco military model features blade of CPM (crucible particle metallurgy) 440V, a tough stainless steel. The black scales are G-10.

your knife is a must—note that you may have to drill a hole through the handle to accommodate a lanyard. Walk away from knives with round handles because in the dark, you won't know which way the edge is facing. Seek a smooth knife handle so that you can use it for hours without feeling like you've been holding a cheese grater. Get a knife that takes a keen edge and holds it—this requires harder steel that takes longer to sharpen. Avoid special-purpose knives because wilderness demands versatility in all outdoor equipment. Manual chain-type saws are lighter and safer than using a hatchet and fit in a nylon belt pouch—buy the kind with links and nylon loops for pulling with gloves on, not a flimsy wire or cable saw. Folding saws also are lighter and safer than a hatchet.

Fire Starters

Most solid and liquid fire starters work well enough. You can buy a few to determine which brand works best before heading into the wilderness. Trioxane has passed the U.S. military's exhaustive testing and is available as surplus—the solid fuel burns clean and hot and isn't messy like gels and liquids. Metal Match also has been tested by the U.S. military and is known as the FS104. It is inexpensive, compact, and it works. An inexpensive BIC lighter or two also will start many fires.

Butane torch lighters are useful, especially in high winds. No one can agree which brand of torch lighter is most reliable or easy to use, so ask campers which brand and model they prefer. If you use gas canister stoves and want something more industrial, consider the Snow Peak GigaPower 2 Way Torch that puts out a roaring 14,000 BTU flame: It comes with an adaptor for CB cartridges and costs $30 to $40. Waterproof matches are a practical addition to every jacket and pack you own. I will not mention primitive fire-starting techniques—hand drills, bow-drills and the like—because they are much better shown on YouTube than described using the written word.

Tents and Tarpaulins

Critical criteria include purpose, climate, season, terrain, number of users, and duration of stay. For example, if you're a recreational camper who only takes the spouse camping on Labor Day in a mild climate, you can buy any number of relatively light-duty tarps or tents. If you're planning a multiweek assault on a significant Himalayan or Alaskan peak, you'll need something else entirely.

Tarps

Tents and tarps are a world unto themselves. Tarps beat tents hands down for versatility and price. Tents beat tarps for applications like the preceding mountaineering examples

Well-worn tarp is hung out to dry, note reinforced corners.

Tarps come in a variety of sizes, colors, and fabrics. Hammer shows scale.

or insect-rich environments where micromesh panels bar entry to airborne invaders. And consider carrying a Space Blanket because it reflects body heat.

Lessons Learned

Quality tarps and tents can serve you and your family for decades, so regard them as investments instead of expenses. Tent failure in the backcountry can be disastrous: High winds or an unpredicted snowstorm can destroy a light-duty fabric shelter so look for strength.

What to Seek and Avoid in Tarps

A tarpaulin's versatility is easily demonstrated: Use it in the morning to cover a load of slash in a pickup bed and in the afternoon (by tossing it over a rope or two) for a spontaneous overnight camping trip. Your application will determine fabric choice. Fabric ranges from water-repellant heavy canvas to silicone-impregnated nylon, which is light and tough (and expensive). You may choose the former to keep in your vehicle and the latter for a backpacking trip. Single-use tarps are ideal for loads that will contaminate or ruin your vehicle: paint, adhesive, animal waste, and so on. The U.S. Army and Marine Corps have been using tarps since the turn of the 20th century. The army calls them shelter halves: Two soldiers, each with a half, can snap them together, hang them over a pole or rope, and stake them to make a pup tent. They can be bought as surplus fairly inexpensively.

Grommets are circular eyelets—usually brass—with good-size holes and are useful for attaching ropes. Well-made tarps will feature reinforcing around grommets for strength and durability. If it's a lightweight fabric, you can gather the center, knot it, and hang it to improvise a shelter. Or you can buy a grommet kit: Grommets can be attached with a hammer or hatchet. To improve ventilation in jackets that don't have pit zips, I put grommets in the underarms.

No grommets? You can create a nondamaging tie-off point by knotting the corners or pushing a smooth pebble or a tuft of grass into the tarp and encircling it with a rope or cord.

What to Seek and Avoid in Tents

Seek stout fabric, tough thread, and high stitches per inch. Look for premium tent poles, because the aluminum alloy or fiberglass usually is of higher quality and lasts longer. Quality pegs are made of aircraft aluminum alloy or high-impact plastic so they won't bend or break at the worst possible time. A blizzard may force you to spend more time in a tent than intended, so buy the best that you can afford. Ultralight tents have become possible because of new designs, fabrics and manufacturing techniques. They're worth considering

Author testing tent near his home in Colorado.

because their light weight enables you to enjoy the journey more. But many ultralights won't stand up to sustained use so look for durability as well as light weight. And *always* remove your footwear when entering a tent with a floor: Lug soles can destroy a coated floor's waterproofing.

Used Tents

Excellent used tents are often for sale, so if you're on a budget or like bargains, look for ads that claim "used-once." New campers often discover that the reality of camping, especially during inclement weather, is more than they bargained for. When buying used tents, make sure to inspect the wands closely, as a bent pole is a warning sign that the owner or Mother Nature abused it. Avoid fiberglass (composite) wands that have begun to delaminate. Inspect coated

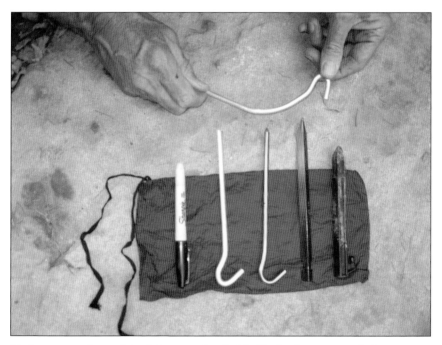

Tent pegs come in a variety of shapes and materials. The bent peg was made from inexpensive aluminum. Marker shows scale.

fabric to see if the waterproof coating has been worn through. Scrutinize seams. Are they coated to make them waterproof?

Newer premium tents and garments feature ultrasonically welded seams, which is advantageous, as the fabric is not pierced by thread. Make sure the insect mesh is intact. If you've spent a night listening to and being bitten by a mosquito, you know the importance of good screens. One or two bent pegs is usual, but if they're all bent, then something is wrong. Pitch the tent to see how long it takes. Can you pitch it wearing gloves? If not, you may want to look elsewhere. Decades ago I owned an expensive mountaineering tent that was made with such close tolerances that I swore every time I pitched it. I sold it at a loss without regret. So be discriminating when inspecting a used tent, with care it will be your safe haven for years.

Mountaineering Tents

While on a backpacking trip on the Continental Divide Trail in Colorado, I shared a friend's Bibler, a lightweight, single-wall (no fly needed) tent. Because of the trail's high elevation, we were hit with daily thunderstorms. The tent's biggest benefit was that *we could crawl inside to pitch it,* so we rarely got soaked, an important advantage.

Look for a tent with flexible wands (poles), which bend a bit to help spill the wind; an important feature for mountain use. Choose a light color fabric in case you get weathered in. Yellow is excellent at providing a cheery interior during days of rain. If your backpack isn't waterproof, be sure to buy a tent large enough to accommodate your pack and keep it dry. If you remember nothing else from this book, make sure to always lie in the tent with the number of people it's rated for before buying. You may discover that a tent rated for three is barely adequate for two adults and their packs or dogs.

Glenn Randall

Author eating breakfast on the Continental Divide Trail under Ute Peak. Yellow Bibler tent got high marks.

Make sure you buy enough tent.

Wall Tents

Wall or cabin tents, which originated in the military, are popular with hunters and other outdoorsmen who tend to spend more than a weekend at a time in a tent. Wall tents are the opposite of ultralight tents and use steel, titanium, or aluminum frames—handy types can save money by building a frame from galvanized-steel electrical conduit. Wall tents are square or rectangular and made from a thick, treated (to resist mildew, shed rain, and retard flame) canvas. Avoid tents made from untreated canvas, as the fabric shrinks after becoming wet, often forcing the owner to cut the frame for easier pitching.

Backpacking tents reduce weight by minimizing interior dimensions, with users crawling in and out. Wall tents feature generous dimensions enabling tall campers to stand under the ridgepole. They come with stove flaps to accommodate different size stovepipes. Wall tents are a great way for new campers to experience the great outdoors, especially children. If you're seeking comfort and roominess and drive to your campsite, they're a solution. Some manufacturers provide a videotape to facilitate pitching the heavyweight home away from home.

Stove brands and types are numerous, so choices are abundant. Some manufacturers offer discounted "camp packages" that include a tent, frame, and stove (most burn wood). And be sure to factor in shipping costs, which can be significant for larger (100-pound) models. Prices range from about $400 to $1,500. Visit websites like outfitterwarehouse.com and walltentshop.com to begin your search, as they offer different sizes from some of the best-known tentmakers. The latter website contains a great deal of useful information and defines terms.

Evolving Tent Designs

Variations of the preceding tent types reach the market as new materials and manufacturing techniques ripple through the industry. For example, Kifaru (kifaru.net) makes several sizes of high-tech tipi (teepee). Tarptent (tarptent.com) makes ultralight shelters.

When tent hunting, determine the features you need and create a list of criteria, including type, size, weight, and price range. Rent or borrow different kinds of tents to see which works for you and which doesn't. Take the same time and deliberation that you would when choosing a vehicle and you won't experience buyer's remorse.

J. Davis

Kifaru's eight-person tipi has no floor so you can wear boots while entering and exiting. Spills are not a problem.

Water Filters

Taking a chance on becoming ill with giardiasis from drinking from what appears to be a clean, clear water source makes no sense. A week after drinking from a tainted water source, giardia can incapacitate you with bloating, abdominal pain, diarrhea, nausea, vomiting, or fatigue. My dogs drink from streams and have become ill from giardia: The smell of their gas and poop is enough to gag a maggot.

Drinkable water provides independence by increasing the distance you can travel while remaining hydrated. Ask experienced outdoorsmen which brand and model of water filter they use and why. Backpack-able filters are of the pump or drip type. Most pumps filter faster than most drips because force beats gravity. My Katadyn pump filter is still going strong after years of use. Mountain Safety Research makes reliable water filters. Camelback offers the All Clear, a battery-operated ultraviolet water purification bottle. According to Camelback: Shake it a couple of times and it provides drinkable water in about one minute—no pumping. Can it get water out of a seep or shallow puddle? I don't know. Katadyn makes a similar product called MyBottle at half the price that uses no batteries, pump, or hose. According to Katadyn, it's the only EPA-registered bottle purification system available. Bottle systems are better for tourist travel while filter pumps excel in the backcountry. If you carry your water in a bladder, be sure to get an adaptor that enables you to pump filtered water directly into it. Bladders are vulnerable to cactus spines so beware.

GPS Devices

Backing up *any* electronic device with a nonelectronic one is prudent. GPS gear is great unless it breaks, in which case you're lost. A map and compass are light, compact, and can save the day. If you use a GPS device in your vehicle, use the same brand when camping. Using different brands means extra work learning a new system. And make sure to update downloadable maps and data at regular intervals.

Sleeping Bags and Pads

Like mattresses, sleeping bags and pads are judged subjectively, so my recommendations will be general. Your optimum sleeping system depends on climate, season, humidity, wind and elevation, and whether you sleep warm or cool or wear clothing or a hat while sleeping. Inflatable pads are great unless punctured. Once punctured, they become ground cloths with little insulating or cushioning capability. Consider carrying a repair kit or preventing punctures by using a closed-cell foam pad underneath to protect an inflatable pad from sharp objects.

Down bags provide the largest comfort range but when wet can become soggy clumps of feathers that refuse to dry. The best way to dry one is in a dryer on low heat (toss in a tennis shoe or two to break up clumps). A synthetic bag is bulky but can be wrung out until damp and dried by sleeping in it. Full-length zippers expand comfort range and zip-together bags are a must for couples because a warm person can warm someone with hypothermia. If you can afford a spring and fall bag as well as a winter bag, make sure to do so. If your budget only permits for one bag, then choose the one that will work in spring and fall, but will be warm enough in winter if you wear clothing and a hat while sleeping. Another option is to carry a waterproof-breathable sleeping bag cover—I opted for a used U.S. military one—that increases warmth while shielding the sleeper from rain and snow. Yes, it's ugly camouflage.

Cookware

Traveling light means one pan that boils water for coffee, fries eggs, and simmers stews. Inexpensive aluminum cookware is plentiful—hard-anodized aluminum requires less maintenance—while stainless steel is only slightly heavier and lasts for decades instead of years. Titanium cookware is as lightweight as aluminum and as strong as steel and is more expensive than either. Pots and pans with copper bottoms tend to reduce hot spots. Coated cookware is an option but is less tolerant of abuse, like using sand to clean it. Whatever you choose, make sure it comes with a good-fitting top.

Rain Gear

The old saying that there's no such thing as bad weather, only inadequate clothing, is mostly true. Quality breathable raingear can keep you from feeling like a drowned rat. But when humping a heavy backpack uphill, you'll be as wet from sweat as from rain. Pit-zips increase ventilation and are recommended. More expensive rain gear stretches with movement, an advantage.

Author setting out on a day hike near his home. Note vintage Lowe daypack and rain-shedding parka and hat.

Compasses and Whistles

Wear a compass and loud whistle around your neck so that you cannot lose them. A compass can keep you from walking in circles when you're exhausted, hungry, and dehydrated. A compass with a mirror enables you to shoot a back bearing (azimuth), signal by reflecting the sun and locate a foreign object in your eye so that you can remove the object. A whistle enables you to signal for hours without getting hoarse and losing your voice. See Chapter Eleven for specific recommendations about winter clothing.

Snowshoeing in deep powder. Note whistle dangling from pack strap.

8

PANDEMIC

For the first time in history we can track the evolution of a pandemic in real time. Influenza viruses are notorious for their rapid mutation and unpredictable behaviour.
—Margaret Chan, Director-General of the
World Health Organization

Certain bacterial infections now defy all antibiotics.
—Stuart Levy, M.D.

Growing global commerce increases international travel. Increased international travel plus antibiotic-resistant bacteria and rapid virus mutation raise the probability that a pandemic can spread faster than epidemiologists can track and isolate it. According to the American Red Cross, an epidemic is the rapid spread of a disease that affects many people in a region or community at the same time; a pandemic is a disease that affects large numbers of people throughout the world and spreads rapidly.

In 1918 and 1919, the Spanish influenza pandemic killed more people—between 20 million and 40 million—than all who died in

World War I. The recent history of delays in determining the cause and reproduction of HIV and AIDS demonstrates that millions of people suffered and died before the world-class Centers for Disease Control and Prevention (cdc.gov) could develop a strategy to prevent it.

The efforts of research physicians and epidemiologists remain frustrated by the ability of viruses to combine genetic elements between strains to create mutations that are both highly contagious as well as lethal. So how do you prepare against such dangerous organisms?

What Does the CDC Recommend?

The CDC (emergency.cdc.gov/agent/agentlist.asp) lists dozens of bioterrorism agents (diseases) alphabetically, from anthrax to Yersinia pestis (plague). The website also has preparedness information for topics unrelated to pandemic and too comprehensive to list here. A downloadable, thirty-four-page "Bioterrorism Readiness Plan: A Template for Healthcare Facilities," dated 1999, lists four diseases: anthrax, botulism, plague, and smallpox. The Isolation Precautions section includes Standard Precautions: Hand washing; gloves; masks/eye protection or face shields; and gowns; followed by patient placement; patient transport; cleaning, disinfection, and sterilization of equipment and environment; and discharge management. No references are made to the gee-whiz biohazard suits with self-contained breathing apparatus or the elaborate airlocks that film and television directors show us. The report does mention isolation rooms with negative air pressure and filtered ventilation. The report is worth reviewing so that you know what to expect and what not to expect from hospitals and clinics.

What Does the American Medical Association Recommend?

As you might imagine, AMA courses are designed for hospital staff and Emergency Medical Services personnel. The "Bioterrorism: Guidelines for Medical and Public Health Management" course provides information about diagnosis and management of infections caused by Category A agents: tularemia, anthrax, botulinum toxin, bubonic plague, smallpox, and viral hemorrhagic fevers. A CD-ROM course entitled "Pandemic Influenza: A primer and resource guide for physicians and other health professionals" also targets healthcare professionals. So look elsewhere for practical information designed for a head of household.

What Does the Red Cross Recommend?

The American Red Cross (redcross.org) recommends roughly the same things that the CDC does and refers website visitors to the CDC website. The flu prevention section lists the same common-sense recommendations that you'll see under "Plans for Pandemics Abound" in this chapter. And the Red Cross sells various kits for informing others about what to do, including a Masters of Disaster program.

The World Health Organization (WHO)

The World Health Organization developed a six-stage classification scale that describes the process by which a novel influenza virus moves from the initial infections in humans through to a pandemic. This begins with the virus mostly infecting animals, with a few cases where animals infect people, then moves through the stage where the virus begins to spread directly among people, and ends in a pandemic when infections from the new virus have spread worldwide. I recall my interest in the news increasing when the WHO number for the H_1N_1 virus zoomed from 4 to 5 to 6 in 2009. Yet relatively few people died of a flu that scored a 6 (pandemic) on the WHO scale, something to keep in mind. Inserting "pandemic preparedness plan" in a search engine yields scores of city and state plans for hospitals dealing with a prospective influenza pandemic.

An Unorthodox Recommendation

While in my early twenties, I succumbed to dengue fever and accompanying delirium as a soldier serving in Vietnam. If you have any thought of "toughing it out" or functioning through any kind of tropical fever, forget it. Imagine my disbelief when I awoke to discover that two days had elapsed since lying on a cot for a quick nap. If a pandemic does occur, my experience as a patient in several military and civilian hospitals makes me inclined to travel in the opposite direction of hospitals to avoid becoming infected.

When a pandemic is first suspected, taking your loved ones camping at a remote campsite probably is safer than remaining in a city or suburb filled with coughing, wheezing, and spitting victims. And because influenza viruses don't respond to antibiotics or some antiviral medications, the treatment you receive at a hospital or clinic will be of dubious value.[4]

Plans for Pandemics Abound

The problem with North American pandemic plans becomes clear if you envision millions of desperate pandemic victims jamming hospitals. It's not the planners' fault that plans cannot prepare hospital-based health professionals for managing a virtual army of victims seeking relief from aching bones, sore muscles, and spiking fevers. Imagine a Hurricane Katrina-like situation—the critically ill filling sports arenas in every state and province. The problem is profound. How many emergency personnel will succumb to the disease or stay home to care for their families? If you see only one film about such a scenario, see *Contagion*.

While doing research about how to prevent pandemic infection, I sought the latest information for my readers but kept coming across the same common-sense preventive measures:

---

[4] Please note that this is the author's opinion and does not necessarily reflect that of the licensee or publisher.

- Cover your nose and mouth with a tissue when you cough or sneeze. Throw the tissue in the trash after you use it.
- Wash your hands often with soap and water, especially after you cough or sneeze. Alcohol-based hand cleaners are also effective.
- Avoid close contact with sick people. If you get sick with influenza, CDC recommends that you stay home from work or school and limit contact with others to keep from infecting them.
- Avoid touching your eyes, nose, or mouth because germs can spread that way.
- Wear an N-95 mask to prevent spreading airborne pathogens.

Decontamination

If you think you've been exposed to a pandemic pathogen, undress, place the clothing you were wearing during exposure in a biohazard disposal container, and shower using soap as soon as possible, then put on fresh clothing.

Pandemic as Cover for Tyranny

If a tyrant wanted an excuse to declare martial law, what better deception could there be than bioterrorism or pandemic? Most of the population would comply immediately with travel restrictions and curfews if they thought it would help quarantine a deadly pathogen. While the preceding scenario is far-fetched it is worth bearing in mind especially if the sluggish economy, political polarization, and cultural decline in America continue.

The Bottom Line

No magic bullet or hold-the-presses pandemic remedy exists. The winner of the race between medical science on one side and drug-resistant bacteria and rapidly mutating viruses on the other has yet to be determined. Medical science has spent billions of dollars worldwide to develop vaccines for both the Ebola and HIV AIDS viruses, but without success. Reviewing what the authorities have done in the past and what they recommend when a pandemic is suspected—for example, the H_1N_1 (swine flu) and H_5N_1 (bird flu) virus—will provide clues to the direction they will take in the future. You may recall that the origin of both the preceding viruses kept evolving as new information became available.

So How Can I Prepare My Family?

What can you do to prepare yourself and your family from the next epidemic or pandemic? My answer is to prepare your immune system against pathogens. The reason that children and the elderly are often the first to fall ill during a pandemic is that their immune systems either are not fully developed or are compromised by the aging process. So strengthening your immune system is critical, and it's fairly straightforward—so straightforward that most people know it but don't do it.

Three Keys to Strengthening Your Immune System

Exercise eliminates toxins, tones and builds muscles, and improves stamina. Exercise also can stimulate the pituitary gland and the hypothalamus to create endorphins, which cause a feeling of well-being. Sufficient exercise also can enable you to fall asleep faster, sleep longer, and more soundly.

Diet can be improved by taking nutritional supplements: Vitamin D3 is particularly helpful at improving the immune system. Because overdosing on oil-soluble vitamins can be problematic, consult a healthcare professional to determine the optimal balance of nutritional supplements for your needs. A comprehensive blood test can reveal which vitamins, minerals, or hormones you are deficient in.

Sleep enables the body to repair itself and the mind to be refreshed and alert in the morning. Sleep deprivation has become chronic in North America so make sleep the priority it should be. One of the first things that skilled interrogators do with prisoners of war is to deprive them of sleep, which demonstrates how important sleep is.

Smoking compromises lung function; alcohol compromises liver function; and illicit drugs compromise brain function, in fact; they can change your brain's chemistry permanently.

Antiviral medications and immune-system boosters are ongoing, with new treatments being developed often. Make sure you pay close attention to announcements by large pharmaceutical companies and new ads for over-the-counter herbal remedies reach the market sooner, as they needn't go through the lengthy FDA testing-and-approval process. With them not claiming to cure the flu, but rather sold as nutritional supplements, they can get on the market quickly.

Undramatic but Effective

The preceding three essentials to strengthening your immune system probably are not the sexy, cutting-edge solution to a pandemic that you may have been seeking. They are, however, what my family and I are doing to prepare from the inside out.

9

NUCLEAR CATASTROPHE

I know not with what weapons World War III will be fought, but World War IV will be fought with sticks and stones.
—Albert Einstein

It's the end of the world as we know it, and I feel fine.
—R.E.M., *It's the End of the World as We Know It*

Einstein was a realist. All-out nuclear war probably would be a world ender, but a tactical (relatively small) nuclear detonation could be survived depending on your distance from ground zero. Currently most people view nuclear war as unthinkable. This attitude can be described as: If they don't think it can happen, it won't. This approach is that of the psychological ostrich. These ostriches would not have believed that, decades after being obliterated, Hiroshima and Nagasaki would be beautiful, thriving cities. While not the result of enemy action, catastrophic nuclear power plant accidents such as those at Chernobyl and Fukushima are lethal events that killed plant workers, rescuers, and affected millions.

This chapter covers practical information for those wanting to know how to survive. Because nuclear weapons, their cascading aftereffects, and your necessary countermeasures are complicated, do your own research on the topic so that you can make a fact-based decision regarding your preparations.

Types of Nuclear and Radiological Weapons and How They Work

A nuclear weapon (atom bomb) is the result of fission in a device containing two or more subcritical masses of uranium. Nuclear fission occurs when the subcritical masses are combined to reach critical mass: splitting heavy uranium atoms, releasing a colossal amount of energy instantly.

A thermonuclear weapon (hydrogen bomb) is more complex, because it involves fusion. In the process, two unimaginably light hydrogen atoms fuse to become one heavier atom of helium. A fission bomb triggers an H-bomb by creating sufficient heat (millions of degrees) to cause fusion and the resulting three-fold energy increase of a fission device containing a similar amount of uranium.

A neutron bomb is a hydrogen bomb designed specifically to release most of its energy as neutron radiation rather than explosive energy. While blast and heat are not eliminated, it is the enhanced radiation that is intended to cause casualties. The neutron radiation produced can penetrate dense materials such as bunker walls.

A dirty bomb is a radiological weapon that uses conventional explosives to spread radioactive material, contaminating the bombsite with radioactive debris. A dirty bomb is particularly dangerous because it requires little technical skill to assemble one. In 2007 a federal jury found Jose Padilla guilty of conspiring to kill people in an overseas jihad. He was accused of building a dirty bomb. Your preparations for surviving a radiological weapon should be similar to those for a nuclear attack.

Electromagnetic Pulse

A side effect of nuclear or thermonuclear detonation is electromagnetic pulse. You will not receive an alert from authorities after detonation because EMP instantaneously destroys most transistorized electronics—though some tube-type electronics will prevail. You'll discover more about EMP and how to defend against it in the following chapter. While deadly to transistors and microprocessors, EMP is not known to harm living things directly.

What Happens Next?

What you need to know is that the chain reaction in a nuclear weapon instantly releases a colossal amount of energy in the form of a fireball—millions of degrees at ground zero—and shockwaves. Radiation given off during the first ninety seconds equals what you'd receive if you were unsheltered for a week—the first week's radiation equals what you'd receive by being in the same location for a lifetime.

Anyone or anything close to ground zero will be incinerated by the heat or destroyed by the blast. (A table listing shockwave effects, including radius [distance] and arrival time for 1-megaton and 10-megaton warheads can be found on page 24 of *Life After Doomsday* by Bruce D. Clayton, Ph.D. While the book was published in 1980, much of the information remains accurate.)

Detonation Location Matters

Where a nuclear bomb detonates certainly matters to you and your family. . . . Especially if your home is near or downwind of a missile silo or U.S. Air Force Base, as either of the preceding increase the probability of being first targets in a conventional war. If a terrorist acquires a nuclear weapon, it probably will be used on a major city. Note nearby likely targets and the direction of prevailing winds so that you can determine primary, secondary, and tertiary escape routes.

A groundburst creates more radioactive fallout than an airburst, as the fireball rises like a hot-air balloon—think of the mushroom cloud as it forms. The rising superheated air creates a low-pressure area that sucks up everything from dust particles to whatever debris the shockwave and fireball did not vaporize. These particles—called radioactive fallout—contaminate anything they touch, wherever the winds take them. Most fallout will dissipate in two weeks, depending on rainfall.

Depending on altitude, an airburst creates less radioactive fallout because the fireball does not create a low-pressure area close to the earth where it can suck up particulate matter. While the lack of fallout is the upside, the downside is that an airburst releases a line-of-sight EMP that can instantly fry the computer on which you store your survival information, so make sure to store valuable survival-related information in nonvolatile form: hard copy, CDs, DVDs, USBs, and the like. Your expensive, multiband radio will become useless if it's not protected in a specific way.

What to Do When Given Warning of a Nuclear Attack

Paying attention to geopolitical events yields many dividends, not least of which is giving you precious time to think and plan as tensions turn to threats and threats to actions. The Middle East, for example, is becoming more and more of a tinderbox. If China or Russia begins ramping up their rhetoric toward the United States or vice versa, pay closer attention to world events.

The Emergency Alert System, under the Public Safety and Homeland Security Bureau, under the Federal Communications Commission (transition.fcc.gov/pshs/services/eas/) provides warning. According to the government website,

> The EAS is a national public warning system that requires broadcasters, cable television systems, wireless cable systems, satellite digital audio radio service . . . providers, and direct

broadcast satellite . . . providers to provide the communications capability to the President to address the American public during a national emergency.

Will the EAS function after one or more nuclear detonations in the continental United States? No one knows.

Rhetoric can turn into saber rattling in a matter of days or hours, so heighten your state of vigilance as events dictate. If a private plane is your primary means of escaping your region, you'd best think again. Planes are too delicate to be trusted after the heat, blast, overpressure, and EMP have their way with them. Even robust military planes will have to be thoroughly inspected before they can be flown. Make certain that wherever you're going is less of a target than where you are. If you have a remote retreat, pack what's critical according to your emergency list and leave. If you have relatives or friends in a safer area, grab your crucial items and call your relatives from your vehicle saying you're on your way. It's better to be wrong about timing than dead wrong about timing.

What to Do Without Warning of a Nuclear Attack

The attack on the World Trade Center towers came without warning on September 11, 2001. The attack on Pearl Harbor also came without warning, so you may get no warning of a nuclear or radiological attack. If you see a flash as bright as the sun and survive the first terrifying seconds of a nuclear detonation, you then have a chance to continue living.

If you're on foot, head for a basement and stay near a corner. If you're driving and have a gas mask with you do *not* stop to put it on: Your immediate threat is from heat and blast, not fallout. Look for a tunnel to drive into or a culvert to crawl into. If you're in a boat or on a ship, unfortunately for you, there's no place to hide. If you're on public transportation, pull or push the emergency stop signal or tell the driver to stop immediately; then exit and seek underground shelter. If you're in open country, take immediate shelter in a ditch or streambed and wait until the shockwave has passed over you.

Goods and Services That Will Be Hard to Get

Whether or not you receive a warning of impending nuclear attack, almost everything will be hard to come by. The next chapter lists important goods that the prudent will stockpile in advance.

Communicating Immediately After a Nuclear Attack

An EMP from a high-altitude airburst can obliterate your cell phone, home phone, and email. Likewise, your television and radio won't work either, so getting accurate information will be nearly impossible. Rumors probably will be your only information source, which means they will be unreliable, speculative, exaggerated, and inaccurate. Establish primary and secondary rallying points for meeting after a nuclear attack (or other major disasters) so that you, your family and nearby friends have a place to meet so that you can exchange nonhysterical information.

Worst-Case Scenario

If all nuclear and thermonuclear weapons were unleashed, experts claim that life on Earth would cease to exist. So you need not prepare for all-out nuclear war. But what if a tactical (smaller) nuke were detonated somewhere? What then? By remembering the information in this chapter you will be less psychologically traumatized following a nuclear attack. When you learn the location of the nuclear or radiological event and the wind direction you can move away from the contaminated area.

Radiation Detection

Because radiation is colorless, odorless, tasteless, and painless, you need a reliable way to detect its presence after a nuclear or radiological attack. While radiation is invisible, radioactive fallout looks like dust or grit.

Most radiation detectors will signal the presence of alpha, beta, and gamma (X-ray) radiation. Alpha radiation consists of relatively heavy, fast-moving, high-energy helium nuclei that cannot penetrate clothing or a piece of paper. Beta radiation consists of light, short-range ejected electrons, for example, strontium-90. Gamma radiation is highly penetrating electromagnetic radiation that is most dangerous to humans. Dense materials like lead can block penetrating gamma radiation. The names for radiation and radiation absorbed by humans have changed over the years and are likely to change again. Read the manual that accompanies your radiation detector and highlight lethal dose and duration, whatever the unit of measure.

You can buy a consumer dosimeter (determines the amount of radiation it has received over time) for $5 or a consumer radiation detector for up to $740 or any price in between. More robust and sensitive models can cost even more. If you want to know only when you are in danger, consider something that fits on your keychain: The NukAlert sells for about $150 and, according to the manufacturer, is a meter, monitor and alarm. The benefit to this type of product is that you usually have your keychain with you. A radiation detector in your desk drawer or attic does you little good.

How Radiation Enters the Food Chain

Immediately following nuclear detonation, radioactive fallout will be everywhere downwind of ground zero. To avoid being contaminated by the fallout, stay in a shelter for the first forty-eight hours. After a week, you can leave the shelter for a half hour for the first week and an hour per day for the second week. After two weeks you can remain outside safely, especially if rain has fallen.

How can you protect yourself and family from radioactive fallout? If you inhale or ingest radioactive fallout, it can make you temporarily ill or can cause cancer later in life. Radiation can create more than 800 radioactive isotopes of naturally occurring elements. Radioactive isotopes in fallout will enter the food chain over time. For example, fallout containing strontium-90—a radioactive isotope of naturally occurring strontium—lands in a pasture; cows eat the grass; going forward their milk is contaminated with strontium-90, which is known as a bone seeker. So by drinking contaminated milk, you are consuming concentrated fallout increasing the chances of contracting leukemia, a form of cancer.

Decontamination

Avoid eating anything that might be contaminated with fallout. Eat what's in your house and wash fruits and vegetables thoroughly in a diluted mixture of water and vinegar. If you're traveling during a nuclear attack or radiological event (dirty bomb), avoid drinking milk or eating fruits and vegetables, and bathe when possible after the detonation, keeping water out of your eyes, nose, and mouth.

Fallout in your hair can drop into the food you're eating, contaminate it, and make you ill. So taking showers or baths to decontaminate yourself, especially your hair, is important. If running water is not available, use club soda or water from a toilet tank, as it's better than not decontaminating. Ceramic or vinyl tile or hardwood floors are easier to decontaminate than those with wall-to-wall carpeting. And a concrete patio is easier to decontaminate than artificial turf.

Protect Thyroid Glands Against Radioactive Iodine

After a nuclear attack, your thyroid gland is vulnerable, as it can absorb radioactive iodine, which eventually can cause thyroid cancer. When emergency management officials or public health officials warn of or announce a nuclear attack, take potassium iodide or potassium iodate to prevent your thyroid gland from absorbing dangerous radioactive iodine-131, a gamma emitter. If you learn of a nuclear attack by other means, don't wait to be told, take the tablets. Buy enough potassium iodide or iodate tablets to last you and your family for several weeks. Potassium iodide is bitter and rough on the stomach, so take it with food or drink. Take the dosage recommended on the container.

Carry the tablets with you and hope that you don't need to take them. I carry some in an inexpensive container that attaches to a keychain. Children may vomit after taking potassium iodide, so give it to them with fruit syrup or a soft drink . . . but get them to take it. Don't wait for a government alert to take your tablets because an alert won't come if an EMP destroys all electronics in your area. You won't be able to buy more potassium iodide or iodate after a nuclear detonation, as your vehicle may become an EMP casualty; so make sure to buy enough the first time. Because potassium iodide and iodate are mineral salts, they have an indefinite shelf life. Potassium iodide cannot prevent radioactive iodine from entering the body, but can protect the thyroid from radioactive iodine. It does not protect other parts of the body and cannot reverse damage to the thyroid caused by radioactive iodine once damage has occurred. If radioactive iodine is not present, taking potassium iodide is not protective.

Gas Masks

Buying a *recently manufactured* gas mask is a good investment, as it can protect you from inhaling radioactive particles. (In August of 2012, I visited the FEMA website (ready.gov) and entered "gas mask" in the search box, but received no information.) The recent masks that the U.S. military and NATO use are excellent, as they protect against chemical, biological, radiological, and nuclear (CBRN) contamination. Seek the most recent NATO surplus mask (they use interchangeable 40 mm threaded filter canisters) and get a mask for each family member. Test a mask for fit (seal)—similar to the way you would test a skin-diving mask—by covering the intake valve with your hand and inhaling for a few seconds. If the mask leaks any air, it does not fit properly. A beard will interfere with getting a good seal between mask and face: Smearing petroleum jelly on a beard will enable a better seal. Like any technological piece of equipment, as new gas masks become available they should be reviewed periodically to see if they are more effective. The following information about gas masks provides a place to start your research; it is not a recommendation.

Avoid Russian gas masks, as most are now obsolete. After producing tens of thousands of defective gas masks (M69, C3, and C4), Canada has yet to field an effective model that is available to the public. If you are considering Israel's M15, make sure that it's of recent manufacture—2003 is a good place to start—don't let the low price of older masks (about $20) influence your decision. And use only filter canisters in current production. Choose a mask rated for CBRN, not one that protects against tear gas.

SGE 400/3 civilian masks get great reviews, accept 40 mm NATO filter canisters, and are approved by the National Institute of Occupational Safety and Health. ProKI (proki.org/sgemask-all.htm) and Approved Gas Masks (approvedgasmasks.com/sge-400-3.htm) had the best price: $143.50. SGE also makes masks for children and smaller adults.

How to Build a Shelter Against Radioactive Fallout

During the Cold War, those Americans fearing nuclear attack from the Soviet Union began building fallout shelters in the 1950s. While the danger of a nuclear attack from Russia has diminished, the danger from dedicated terrorist groups has increased. Because of this, building a safe room can still serve several purposes: It can be a refuge from a home invasion, hurricane-force winds and tornadoes, and radioactive fallout.

Shelters are available at several websites. American Bomb Shelter sells modular components, kits, and plans for bomb shelters, so you can choose only the items you or your general contractor think appropriate for your project. For example, installing a filtered ventilation system will prevent airborne radioactive fallout from contaminating your safe room. To see an example of a filtered ventilation system visit americanbombshelter.com/ASR-50-AB-automatic-bunker-ventilation-blower.htm

How and Why to Build a Safe Room

To build the best possible safe room, excavate some distance from your residence to ensure that the structure cannot collapse onto the safe room exit, trapping those inside. Building a safe room in a basement is much more cost effective. If your house has no basement, do the best you can while understanding that it will be suboptimal: Some handgun and rifle bullets can penetrate both exterior walls of a frame house and continue on to wound or kill someone on the far side. Basement walls, which usually are poured-in-place reinforced concrete, are a great place to start. A basement corner is a particularly advantageous location for building a safe room because two walls already exist. If your basement has an area bounded by three walls, all you need to do is build the fourth wall incorporating a door.

I planned our safe room by using masking tape to lay out the space. It will accommodate three or four people in a pinch. In the fall of 2008, when the stock market fell off a cliff, I built the safe room for my family in case of civil unrest. I take satisfaction in knowing that no one has guessed it's there. If you aren't comfortable building a safe room, several companies will build you one ranging from stark and simple to large and lavish. While a safe room will protect your family from radioactive fallout, it will not withstand a nearby nuclear shockwave: The crushing overpressure destroys it. A safe room also can be used as a strong room in which to store your valuables including water and food, and weapons and ammunition.

If you live in an aboveground residence, consider partnering with a family you know and trust. You can financially or physically assist a landowning, like-minded family to build a safe room large enough to accommodate two families. The other family members need not be friends, but they must be reliable and trustworthy allies in case of a nuclear disaster.

How to Build Safe Room Walls Using Concrete Blocks

When completed, author's safe room looks like it was built as part of the house. Using half-blocks enables staggering the full-size concrete blocks.

Some stores don't carry solid block, which is expensive and worth it. You can use hollow, heavy-wall blocks instead (see table below). Most concrete blocks have a notch for inserting reinforcing bar (rebar). If you are proficient with the masonry trade so much the better. While heavy-wall, hollow block is stout enough for many applications, it needs additional mass to be suitable for a fallout shelter or safe room. To strengthen hollow block, the voids can be filled with sand, pea gravel, or concrete. I chose sand because it flows easily through the staggered voids and because mixing concrete is labor intensive and relatively permanent. Almost a ton of sand went into the wall, which weighed a couple of tons when completed. A concrete basement floor can withstand this kind of load safely. After cleanup, the wall can look like it was built as part of the house.

Structural concrete block comes in five common sizes (actual dimensions are smaller):

Concrete block size	Configuration	Note
4-inch x 8-inch x 16-inch	Solid or hollow	Solid recommended for safe rooms
6-inch x 8-inch x 16-inch	Hollow	
8-inch x 8-inch x 16-inch	Hollow	
10-inch x 8-inch x 16-inch	Hollow	
12-inch x 8-inch x 16-inch	Hollow	

Decades ago, I did enough repair work on a brick wall to know that I am lousy at masonry. For above-grade and non-load-bearing basement applications, an alternative to mixing mortar and applying it with a trowel is applying liquid adhesive with a brush. This method was something I could master.

A chalk line on a concrete floor acts as a guide for the first course (layer of block). Apply the white adhesive liberally to the floor and a few blocks, wait until the adhesive is tacky, and set each block in place. You will have adequate working time before the adhesive sets. As it dries it will become transparent, which makes the finished product look more professional. Use half blocks to stagger the joints for each course (layer). If the first course is straight, subsequent ones also will be straight if you're careful.

Do *not* build such a wall on an upper floor or conventional wooden subfloor, as it certainly would break the floor joists. If the preceding description of building a safe room seems beyond what you can do safely, consult a trustworthy contractor and help to plan it as if your life depends on it. It may.

An Alternative to Masonry

Our safe room was now complete except for a portion of a sheetrock wall with a closet. Not having room for a block wall in a closet, I removed the existing dry wall to expose the studs. Next I drew a sketch—including dimensions—and faxed it to a steel vendor, ordering 3/8-inch mild-steel plate. A friend helped me pick it up and load it on my truck. We muscled the pieces into the room, drilled holes in the steel, and bolted it to the studs. The installed steel plate was later finished in drywall, taped, textured and painted to match the rest of the room. Note: When ordering steel plate, consider having the vendor cut each 4 x 8-foot sheet of 3/8-inch steel plate in half lengthwise. The extra cut will add to the cost, but will facilitate handling it. A full sheet weighs 480 pounds.

Our safe room provides security for my family and loved ones. While we hope that we never need to seek refuge inside it—especially from a nuclear or radiological attack—we're glad it's there. If I'm away from home, my wife and daughters can remain inside while calling authorities, if there are any authorities left.

Defending Your Safe Room or Fallout Shelter

After going to the trouble and expense of building a safe room to protect your loved ones, it stands to reason that you will want to defend it against intruders and looters. Few safe rooms can withstand sustained physical attack by determined and resourceful people, so defending your safe room is a logical link in the chain of preparation. See Chapter Eleven for specifics regarding weapons and ammunition.

10

ELECTROMAGNETIC PULSE (EMP)

Several potential adversaries have or can acquire the capability to attack the United States with a high-altitude nuclear weapon-generated electromagnetic pulse (EMP). A determined adversary can achieve an EMP attack capability without having a high level of sophistication.

EMP is one of a small number of threats that can hold our society at risk of catastrophic consequences. EMP will cover the wide geographic region within line of sight to the nuclear weapon. It has the capability to produce significant damage to critical infrastructures and thus to the very fabric of US society...

—Report of the Commission to Assess
the Threat to the United States from
Electromagnetic Pulse (EMP) Attack
[2004 Executive Report]

U.S. Vulnerability

"Catastrophic consequences" and "critical infrastructures" in the preceding quotation sound the alarm without sounding alarming. The truth is that one or more high-altitude nuclear airbursts over America would cripple most of the power grid, instantly turning the United States into a developing nation.

On July 9, 1962, project Starfish Prime disrupted a city power grid in Hawaii, *900 miles* from the high-altitude (250 miles above sea level) nuclear detonation. Terrorists and nations hostile to the United States are aware of infrastructure vulnerability to EMP and are committed to producing and launching missiles armed with nuclear warheads to cause such a catastrophe.

The previous chapter mentions the devastating effects of EMP caused by a high-altitude nuclear airburst. This chapter describes the massive infrastructure damage, details how to protect your sensitive, critical electronics, and explains which items to stockpile.

How EMP from a Nuclear Weapon Works

A nuclear explosion in the upper atmosphere creates a two-billion-volt pulse knocking loose charged particles and focusing them toward Earth. As the pulse, diminished by the distance traveled from the explosion point, speeds through the atmosphere, it becomes diluted. Wires, cables, pipes, and conductive metallic objects can increase the voltage spike.

All nuclear weapon detonations create an EMP. For ground detonated nukes, the effective EMP distance may only be a few miles, while low airburst weapons may have a range of five to ten miles. The focus of concern for developed nations is a weapon detonated at an altitude between 250 to 300 miles, the part of the atmosphere filled with charged particles. The missile needs only to get the nuke to a general location rather than have the pinpoint accuracy needed to hit a target on the ground.

How an EMP Attack Can Occur

If a substantial nuclear weapon were detonated over Pierre, South Dakota, and Oklahoma City, North America would be reduced to 18th-century technology. Four smaller, freighter-launched missiles with 1,200-mile ranges if geographically dispersed would destroy the operations of North America's major population centers. Iran possesses such missiles. EMP strength from a nuclear weapon is directly dependant on a number of variables including the base power of the weapon, detonation altitude, and the type and shape of shielding around the bomb.

Currently, the Internet provides detailed information about how, during WWII, America's Manhattan Project developed the first atomic bomb. Nuclear technical processes are widely known. Nations and terrorist groups with access to nuclear materials, specific technology, and a sophisticated scientific community can build a nuclear device. Many nuclear scientists were furloughed when the Soviet Union dissolved, and countries hostile to the West likely employed them.

Iran's population has more advanced degrees per capita than any other country. Iran has had an active nuclear program for nearly twenty years and knows the evolutionary process of creating sophisticated nuclear weapons. Iran procures bomb components from a worldwide black market in timers, electronics, altimeters, and sophisticated switches—crucial to controlling conventional explosives to compress subcritical masses of highly enriched uranium to reach critical mass. In 2012, Iran doubled the number of centrifuges—from 1,064 to 2,140—needed to create highly enriched uranium for nuclear weapons. The centrifuges are in operation at Iran's hardened Furdow facility.

Iran has sophisticated Russian-built diesel submarines that are quieter to sonar than American nuclear subs. Iran has run tests to raise their Shahab 3 missiles on hydraulic lifts from the holds of freighters, fuel and launch the 1,000-mile-range missiles in fifteen minutes or less. These missiles could carry nuclear weapons into North America's atmosphere. The ability of the United States to destroy such missiles in flight is doubtful. Iran has been working diligently to make more powerful nuclear weapons that also are smaller, lighter, and more easily configured to fit into a missile.

EMP Produced by the Sun

Earth's sun is an ongoing thermonuclear furnace that produces EMPs, especially during severe solar storms or coronal mass ejections (CME). Imagine the colossal power that can create a sunspot area thirteen times the Earth's diameter, produce heat exceeding 10 million degrees Fahrenheit, and shoot an EMP more than 90 million miles to cause chaos here on Earth. The powerful pulse can destroy sophisticated electronics instantly. If a severe solar storm erupts on the side of the sun facing Earth, because of the 92 million miles between the two, the authorities should have time to alert the public about the impending EMP event. The latest solar storm was in the fall of 2003, and caused hundreds of millions of dollars of damage worldwide. More CMEs are due in 2013, the year of the solar maximum. Almost all electric grids and the homes and businesses connected to them are vulnerable to EMP. A CME also interferes with the network of orbiting satellites that constitute the global positioning system, another reason for concern because the U.S. military relies on it.

Long, Grim Aftermath

The U.S. Government predicts that 80 to 90 percent of the populace would die within twelve to twenty-four months from the effects of starvation, disease, and civil disorder after an EMP attack. The people and knowledge base needed to rebuild the infrastructure probably would be lost. For financial reasons, most electric utilities protect infrastructure against lightning, while leaving it vulnerable to EMP. This makes no sense. The cost of EMP shielding, while expensive, pales in comparison to rebuilding the massive electrical grid. And rebuilding is problematic because each of the huge 350+ power-station transformers is constructed overseas and takes months or years to build. If most of the transformers in the United States were destroyed on the same day, they would take years or decades to make and install . . . a horrible scenario by any standard. Currently, the United States does not have the capability to manufacture large transformers. And without a functioning power grid, industry could not build a facility in which to produce such transformers. Nicholas R. Forstchen's novel, *One Second After,* paints a dark and dismal picture of American life after an EMP strike.

Reliance on Microprocessors and Computers

If you are an information technology or computer science professional, you need not read the remainder of this paragraph, as you already know how pervasive microprocessors and computers have become. Computers and smart devices quietly control many aspects of our lives without us realizing it. Do you take for granted the uninterrupted flow of electricity from the public or private power utility? Most of us have grown up with it and use it daily. What about the microprocessors and computers that you aren't aware of? Do you know about the numerous computers in your late-model car that manage every aspect of your engine? For example, they inform your ignition when to advance or retard engine timing, and instruct your fuel injectors how much fuel to supply. A computer tells your transmission when to upshift and downshift.

How to Protect Your Electronics Against EMPs

Regardless of an EMP source, you must shield critical electronic devices. Any electronics using a microprocessor or integrated circuit are prone to fail if an EMP is powerful enough. Placing spare electronics in metal containers protects them from an EMP's long, lethal reach. Consider unshielded electronics expendable. In general, conductivity beats density for shielding, so copper is ideal for such a container, but high prices make it too expensive to be practical. Aluminum is the next best metal for a container. For smaller electronic devices, use aluminum foil as shielding. By placing the aluminum-foil-wrapped items in a plastic bag and wrapping the bag in another layer of aluminum foil, you can be fairly certain that you've prevented EMP from zapping your electronics. Overlap all foil seams and be sure not to leave gaps or holes in the foil, as they become potential paths of entry for EMPs. If all you have is a trash-can and lid to protect your electronics, it's better than nothing. Use insulation between the can and the object but not between the can and lid.

Wires entering or exiting an EMP-shielded container protecting your items can create an *antenna effect:* Wires can carry the EMP into your container, defeating the purpose of shielding. If your electronic device uses batteries, consider shielding numerous lithium batteries, which have a ten- to fifteen-year shelf life. Include incandescent light bulbs, LEDs, and compact-fluorescent lights, as they will be valuable, especially if you have a solar-powered generator.

Water Will Be Hard to Get

What makes running water run? Valves and pumps maintain your water pressure. What controls the valves and pumps? A programmable logic controller: A small computer that comes in various sizes and configurations. PLCs manage pumping and flow of water and sewage, natural gas lines, petroleum pipelines, electricity distribution and thousands of other applications required for infrastructure to operate. The lines attached to PLCs, the amount of shielding, and so on determine if all or most controllers will be destroyed during an EMP attack. Even if engineers were able to repair or replace power station transformers, the substructure of the distribution network also would need replacement.

Storing water is prudent for city dwellers and those getting water from a well using an electric pump. Water is heavy, so experience teaches that water stored in smaller containers, five to ten gallons, is the maximum to transport by hand.

Food Will Be Hard to Get

Just-in-time inventory has increased the efficiency of distribution. Companies using JIT require less warehouse space, as computerized inventory software enables calculating the best time to deliver the optimum amount of supplies, including food and beverages. If an EMP prevents inventory software from tracking and trucks from transporting, food will become scarce virtually overnight. History demonstrates that hunger often leads to civil unrest and looting as people try to feed their children. If you doubt such behavior can occur in the United States you have only to recall the destruction that happens in large cities when sports teams win major sporting events, or the looting during power outages.

Accurate Information Will Be Hard to Get

A severe solar storm or nuclear-weapon-caused EMP will cause chaos in the upper atmosphere, so it may take hours after an EMP strike before you begin receiving any information on a shortwave band, but it likely will arrive before other information can reach you. A severe solar storm or nuclear-caused EMP could cripple radio transmitters for years. A NOAA Weather Radio station can provide useful emergency information so keep a couple of radios shielded so that you have a reliable backup source.

All Things Medical Will Be Hard to Get

Doctors don't make house calls these days, so imagine what it will be like after an EMP attack or solar storm. Make a list of nurses, physicians, and pharmacists who are neighbors—if they have not prepared so well as you, they'll be more motivated to make house calls in the surprising quiet after electricity stops flowing. Check your emergency medicine supplies and stock up on prescription medication for chronic conditions. Update your cardio-pulmonary resuscitation skills and seek additional emergency medical training so you can protect your loved ones.

Emergency Services Will Be Hard to Get

Firefighters, police, and ambulances can have difficulty reaching a destination during rush hour. Imagine trying to navigate if most vehicles are immobile? And how would emergency agencies communicate with others? You would be wise to buy large fire extinguishers and keep them handy. When law enforcement is unavailable—they have families, too—you become responsible for maintaining law and order. Talk to neighbors about forming a neighborhood watch group. When turbulent times come you'll be part of a team instead of a lonely ranger.

Eyeglasses Will Be Hard to Get

Obtain additional prescription eyeglasses so that you have vital spares. While not having a spare pair of prescription glass is an inconvenience during normal times, it could be life threatening after an EMP has ravaged your community and you can no longer get glasses made.

Which Vehicles Are Immune to an EMP?

EMP Commission testing showed that EMP would permanently damage roughly 15 percent of representative vehicles in the study. Vehicles not using transistorized ignitions and computerized engine-management systems—those made before 1965—are mostly immune to EMP. Pre-1965 ignition systems use a coil, rotor, contact points, and condensers to fire spark plugs and are not affected by EMP. Diesel engines that ignite fuel using high-compression ratios also are thought to be more resistant to EMP than gas engines, but *newer diesels contain numerous computers* to keep emissions within EPA limits. Bear in mind that no real-world EMP testing has been conducted since 1962, so opinions about the effects of EMP abound. In general, U.S. military vehicles contain more shielding than commercial vehicles and are considered immune.

Are Tube-Type Electronics Immune to an EMP?

No, they are not immune, but most electronic devices using vacuum tubes are much more resistant to the effects of an EMP than transistor models. Years ago, Soviet pilots seeking asylum flew their jets to non-Soviet countries. Many people laughed when these state-of-the-art jets were discovered to employ vacuum tubes in their communication gear. Now that you know about the effects of an EMP, you can understand why the Soviets may have used old-fashioned, tube-type transceivers in jets that were otherwise up to date. If you own or are considering getting tube-type devices, know that they are less efficient than their transistorized counterparts, so buy more batteries than you think you'll need.

Off-the-Grid Solar-Powered Systems and EMPs

Deciding to live off the electric grid by using an array of photovoltaic panels to power a home is a great way to become more self-reliant. Some solar companies offer contractual agreements to homeowners that authorize the company to install a solar-powered system and lease it to the homeowner at rates that rival or beat average monthly utility bills. Because large solar arrays can produce more power than needed, a homeowner can link the system to the electrical grid and sell power to the utility company. In case of an EMP attack, however, this link to the grid guarantees that the utility's transmission lines become a highway to destroy the solar-powered system and every vulnerable device and appliance connected to it.

Buy or Build a Portable Solar-Powered Generator

You are prudent to own a gasoline- or diesel-powered generator to use during brief power outages. But after an EMP event, a solar generator rules the day. Why? Because it produces no exhaust noise or smell to signal others that you have power. And you needn't hunt for fuel—it comes to you.

While still in their infancy, portable solar generators represent the future. New products are coming to market as fast as manufacturers can make them. If you live in a region that's chronically overcast, a solar generator will be less useful. If you plan to use your generator in the tropics or arctic or on an ocean-going vessel, buy the most robust unit available. Time spent using Internet search engines and taking good notes yields handsome dividends—it can save you from paying too much, or worse, investing in a sub-par product.

Definition of Key Electrical Terms

Being familiar with technical terms will assist you in wading through generator specifications.

AC—an electric (household) current that reverses direction in a circuit at regular intervals

AMPERE (AMP)—a common measure of electrical current (flow) in a conductor

DC—the continuous flow of electricity (from a battery) through a conductor (wire)

GFCI—ground-fault circuit interrupter, protects in areas where people are likely to receive an electrical shock (in damp or wet locations, for example)

VOLT—a measure of electric potential (think of voltage as water pressure in a pipe)

WATT—a unit of electrical power (for example, light bulbs use watts as a unit of measure). Watts are the standard used by manufacturers to specify a conventional or a solar generator's output power—think of watts as similar to horsepower. Following is a rough rule of thumb for classification.

250 WATTS OR FEWER: Suitable for small household appliances, televisions, DVD players, computers, and small office equipment, and LED lighting. Many connect with a 12-volt plug.

500 WATTS: Suitable for larger household appliances, including large stereos, countertop kitchen appliances, and power tools. Most have two grounded receptacles for powering devices.

1500 WATTS: Suitable for most household appliances, larger power tools, and incandescent lighting. Some systems are on wheels, significantly enhancing their versatility.

Sine Language

If your intended generator use is running power tools in the wilderness you can disregard the rest of the paragraph. If you will use a generator for charging computers, laser printers, fax machines, or other power-sensitive devices, you'll need to investigate models that produce more expensive pure (true) sine wave power—similar to that produced by a public power utility. True-sine inverters cost two to four times as much as those making modified (square) sine-wave power, which is fine for many electrical devices. True sine and modified sine refer to the waveform (shape) of an electrical impulse as seen on an oscilloscope.

How Does It Work?

A portable solar-powered generator does what an internal combustion generator does. It provides power to run household appliances and tools during a power failure or for use in a remote location (off grid). As with an internal combustion generator, you must match the output (in watts) to the generator's intended use. Unlike a gasoline or diesel generator, a portable solar-powered generator stores energy in a battery adding to its versatility.

What's Inside?

Like conventional generators, solar-powered generators contain components that function as a system, including:

- Inverter—changes DC or direct current from a photovoltaic panel (12 or 24VDC) to AC or alternating current (120 or 240VAC) so a user can plug an electrical device into a standard household-type receptacle. Good inverters are 90% efficient. The inverter and photovoltaic panel represent the heart and lungs of a solar generator. More expensive inverters create true sine wave electricity, while most generators create modified sine wave electricity.
- Battery (12VDC)—stores power from a photovoltaic panel(s). Battery type can be lead-acid (automotive type) or no-maintenance AGM (absorbed glass mat). Using a charger—plugging the generator into a 120VAC household receptacle—is another way to charge the battery. When replacing a battery, determine if it is a regular or deep-cycle type. The former is a standard automotive battery. The latter is heavy duty (like those used in golf carts) and withstands more charge-discharge cycles than a standard one.
- Charge controller—monitors and directs DC voltage from a photovoltaic panel to charge the battery. The controller automatically prevents overcharging and damaging the battery. Ask if a charge controller is included in the system or if you need to order one separately.
- Photovoltaic panel—converts sunlight into electrical power through a process occurring at the subatomic level. Photovoltaic cells usually are made with silicon: One side has a surplus of electrons and the other side has a lack of them. This condition generates DC electricity, charging the battery.

- Receptacles—enable the use of automotive-type cigar-lighter plugs or standard household plugs to power DC devices and AC devices respectively.

Assemble Your Own

For do-it-yourselfers, a number of generator kits are available. The advantage of a kit is twofold: You can save money by assembling one, and you'll find it easier to repair something that you've built. The downside is you may not get so good a warranty. Check on warranty coverage if considering a kit. If you're an electrical whiz, you can select and assemble components that best suit your application.

Photos show sequence of author mounting components (Xantrex XPower Powerpack 1500® generator, 100-watt Kyocera® photovoltaic panel, and Morningstar's SunSaver-10 charge controller) on hand truck to assemble a solar-powered generator.

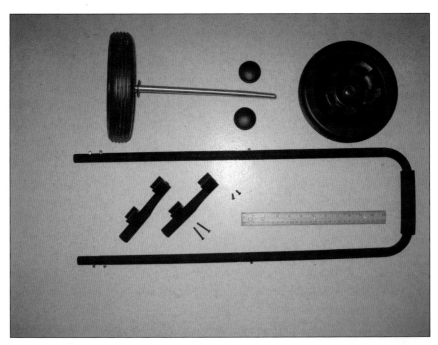

Discarded parts from Xantrex 1500 generator: plastic wheels on metal axle, tow handle, and mounting hardware. Ruler shows scale.

Tightening blue nylon ratchet straps securing generator case to hand truck frame. Note padded aluminum base for photovoltaic panel. Ruler shows scale.

Drilling holes through photovoltaic-panel frame and horizontal aluminum support. Wood wedges hold aluminum support in place during drilling.

Drilling holes in horizontal support to receive U-bolts that fasten photovoltaic panel to hand truck frame.

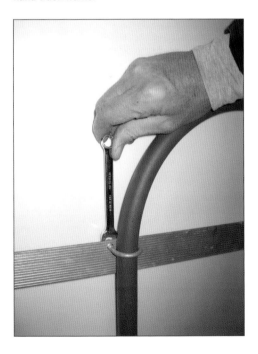

Tightening nut on U-bolt securing panel to hand truck frame. Background is white plastic rear of photovoltaic panel.

Circular clamp fastens strap-iron stand to hand truck handle.

SunSaver-10 charge controller attached to bottom crosspiece of hand truck. Note yellow, weatherproof, heat-shrink tubing on terminals. Background: loops of black wire from rear of PV panel. Illuminated green LED indicates charging.

Using a heavy-duty drill in the field. System is leaning on stand to maximize sunlight on PV panel.

Determining Generator Output

Buying enough generator output is important. For example, if a generator will be tasked with powering a heater, you'll want more output. Weight also matters—if you plan to carry or roll your portable generator to remote locations, make sure the weight is well within your limits or you may be reluctant to use it.

Pre-Purchase Check List for a Solar Generator

Include a generator's output power in watts (either true sine wave or modified sine wave), AC power output in volts and amps, number of grounded receptacles, GFCI capable, battery capacity in amp-hours, battery type (sealed or lead-acid, standard or deep cycle), battery weight, overall weight, overall dimensions, useful features (lifting handles or wheels, for example), country of origin, warranty type and length, after-purchase service and tech support, plus miscellaneous (for example, an audible alarm to warn of low battery charge), and create your own categories. I recommend only solar generators made in North America. Look for a UL (Underwriters Laboratories) label in the U.S. and a CSA (Canadian Standards Association) label in Canada.

Relevant Websites

The following websites are useful for familiarizing yourself with the growing world of solar products. Use search engines to refine and enhance your research. The following list of resources is provided for informational purposes only.

poweredgenerators.com/solar-generators.html
ecogeekliving.com
bhasolar.com
thesolarpowerwiki.com
theinverterstore.com
bluepacificsolar.com
energyrefuge.com/how-to/solar_power_generator.htm

Shielding a Solar Generator from EMP

A solar generator is difficult to shield because photovoltaic panels are large. When not in use, store your solar generator in a safe-room closet that is totally lined with thin aluminum plate or thick aluminum foil, an insulation layer (sheet plastic), and another layer of aluminum.

EMP Expertise Beats Guessing

Protecting your loved ones in a post-EMP environment is not intuitive or inexpensive. If you're going to the trouble and expense of keeping an EMP from jeopardizing your family by annihilating valuable assets, either acquire EMP knowledge or consult an expert.

11

PREPPING TIPS AND TECHNIQUES

When I was a boy, I always saw myself as a hero in comic books and in movies. I grew up believing this dream.

—Elvis Presley

Arms in the hands of individual citizens may be used at individual discretion for the defense of the country, the over-throw of tyranny, or in private self-defense.

—John Adams

While the previous chapters provided information about major disaster scenarios and ways of preparing for and dealing with them to safeguard yourself and family, this chapter contains information applicable to several situations. This chapter:

- Dispels the lone wolf myth and stresses the importance of community
- Explains the value of intelligence
- Provides details about cold weather clothing

- Assists in determining which firearms and knives are best for certain purposes
- Discusses so-called "survival" products
- Provides a few tips about bartering.

The Romantic Myth

Novels, Hollywood movies, and some blogs portray the heroic exploits of solitary men or women. These fictional heroes don't run out of water, food, ammunition, or great gadgets, despite the odds or circumstances. So we can hardly be blamed for devouring such dramatic and unrealistic stories. Most of us have read books and blogs that advocate grabbing a bug-out bag and heading for the hills in times of disaster. But the question is where did this lone wolf idea come from?

John the Baptizer lived off the land eating locusts and honey—a loner if ever there was one. During the years that Jesus Christ was a shepherd, he was a loner. How about Buddha? He reputedly spent years alone meditating and contemplating. The preceding examples remind us that the myth is an ancient one.

We can then fast forward to American history and well-known legend Henry David Thoreau. Thoreau was a loner who communed with and wrote about nature. Daniel Boone and David Crocket spent a great deal of time alone outdoors. Our comic books reveal numerous good or evil loners battling to save or enslave the world. The point is our culture has many versions of loners who prevail in adversity. Whether you call it going solo or being a rugged individual, we grew up with these dramatized tales. Americans love the idea of one person winning against all odds. The fact that most loners don't succeed doesn't seem to stop us from loving the myth.

Getting Realistic

Most significant achievements are made by groups of people striving toward a common goal. Whether space travel, battlefield victories, or Olympic triumphs, it's teams that succeed. For example, U.S. Navy SEAL teams are teams for a reason. Baseball, football, and basketball are called team sports for a reason. So maybe we like the lone wolf myth because it is so unlikely. Most of us cannot relate to a group so we romanticize rugged individuals because it's easier to identify with them. Going solo doesn't work for numerous reasons so let's review a few. No one can:

- Stay alert or awake 24–7
- Be in two places (front and back door) at once
- Provide his or her own cover or back up
- Know the numerous skills required to function during and after a major disaster
- Remain healthy or uninjured all of the time.

Focusing the Search

So if going solo is dangerous at best, how can we protect our loved ones? Finding an ally is neither easy nor can it be done in haste. Getting to know and learning to trust someone takes time. Yet by narrowing the search to people possessing the character traits you want, the task becomes easier. Start by looking for honest, loyal, discrete, moral, and principled people. As a rule, former police, military, and emergency personnel make good prospects. Those with families to protect also are potential allies. Houses of worship, service organizations, charities, volunteer fire and rescue units, and the scouts are excellent places to go prospecting. Shooting ranges also can be useful for meeting like-minded people. (I tend to avoid chest thumpers and back slappers.) Again, the process is neither easy nor fast.

Building a Team

After choosing suitable individuals or families, introduce them to each other to see if they get along. They need not become friends to be solid allies in troubled times, but they need to trust and respect each other. Encourage those with useful skills and knowledge to share them with other team members—this can be as formal or informal as the participants want. Share books, articles, and videos that assist in learning valuable subjects. Bring in local experts to hold workshops on topics that will be useful in case of a disaster: emergency medical care, gardening, food preservation and storage, neighborhood watch groups, non-petroleum-reliant transportation (horses and mules), radio communication, and whatever else is relevant. For example, snow-machine maintenance is appropriate in Alaska, and swamp-buggy maintenance is right for the Everglades. In general it's easier to meet like-minded people in small communities than urban areas.

Time spent working together is the glue that bonds individuals into a team. Not meetings or social gatherings, but time spent training and completing projects to achieve objectives. Americans are famous for pitching in to help neighbors achieve a common goal. For example, barn raisings; farmers helping the sick or elderly to harvest a crop; and forming volunteer bucket brigades to put out a fire in the community. It's common for team size to grow or shrink as families move into or out of a neighborhood or community. What's crucial is quality and commitment, not quantity. And sharing a common faith is proven to knit a team together.

Team Wisdom

A recent book by James Surowiecki, *The Wisdom of Crowds*, states that under the right circumstances, groups are remarkably intelligent and often are smarter than the smartest person in them. Groups don't need to be dominated by exceptionally intelligent people to be smart, and this collective intelligence can be enhanced by group members' diverse backgrounds. So a team has proven reasons to be more successful than a lone wolf, regardless of how talented, cross-trained, and skillful that person is.

Hard Choices

Your vision, purpose, and priorities may not be those of other team members. For example, if your priority is gardening while the others favor marksmanship, you can either learn marksmanship or form another group. As founder, you may have to swallow hard and pass the leadership role to a team member who has more leadership experience. If you find others gravitating and looking to another member for leadership, be honest and gracious enough to invite the other to lead the team. As contradictory as it may seem, to become more self-reliant we have to depend on trusted others. My hope is that the preceding principles encourage you to build a resourceful, resilient, and enduring team.

Charity and Security

After a major disaster, food will be scarce. Your challenge is to balance charity with your family's security, a difficult decision. While you can give the unprepared a bag of food, do not invite unknown persons into your home during times of emergency. You can be polite and charitable without creating awkward or dangerous situations in your home. Keep in mind that before a disaster others possessed the same information that you did, but they chose not to prepare. Discuss the preceding with your family so that they have time to see the wisdom of not being overly hospitable to strangers. Children will have to mature quickly if hunger is the rule and violence is in the wind. Selfless and overly sensitive types also will have to toughen up a bit to function in a harsher reality.

Intelligence Can Be Critical

Accurate timely intelligence from a reliable source always trumps ignorance and second-guessing. Information can be more important than water or food in some cases. For example, if you know that on Friday afternoon the banks will declare a "holiday" and won't open again for several days, you can plan accordingly. If you know that the price of a certain commodity—grain, precious metal, ammunition—is about to spike, you can buy it to hold or as a speculative purchase.

One way to receive such valuable intelligence is to exchange information with a network of reliable people. This network takes time to build, but pays large dividends. Email is a fast and easy way to exchange low-level intelligence. For example, an estate sale will be held in a neighboring community next weekend featuring survival items. For more sensitive information, face-to-face meetings are safest. Another way to receive actionable intelligence is to subscribe to newsletters, which can be laser-focused on a topic or more general. Some newsletter writers will send prospective purchasers a one-time, free back issue of their newsletter. Paying for information may seem odd given the abundance of free Internet information, but knowing specific, time-sensitive information before others is worth paying for.

Government sources may not provide accurate information during an emergency. For example, the information immediately following Hurricane Katrina in 2005 was disjointed and contradictory, which reflected the poor communication and coordination between the state and federal governments. If you lived on the Gulf Coast at the time, you'd want specific, accurate information for short- and long-term planning.

What to Consider When Buying Cold-Weather Gear

Hunters and others spending time outdoors in cold weather know it's critical to stay warm. The alternative increases the possibility of hypothermia, frostbite, or both. Accounts of mountaineers losing fingers and toes to frostbite are commonplace. Armed with knowledge about clothing, you stand a better chance of enjoying time outdoors instead of suffering through it. North America has an enormous climate range, so if you'll be traveling to an unfamiliar region, find out specifics so you can dress adequately.

Field testing Kifaru's two-piece 24–7 CCS™.

Most of the following information comes from personal experience. Trekking Nepal's Himalayas, and a year working north of the Arctic Circle in Prudhoe Bay, Alaska, have provided a deep respect for frigid temperatures and the preparation and vigilance they require. Worst case, inadequate clothing can be deadly.

How to Determine Which Clothing Is Optimal

Temperature, humidity, wind, elevation, and duration outdoors provide a solid foundation for choosing appropriate clothing . . . but there's a lot more to know before dressing and stepping outside. As you'll discover, temperature ranges provided by manufacturers mean little because of the number of variables involved. What will you be doing? Savvy outdoorsmen will wear one kind of clothing when hunting from a tree-stand, and quite another when humping heavy loads up steep hills. Will you be doing the same activity—splitting wood, for example—for hours or will your activity be intermittent? If clouds block the sun, how much does the temperature change? The proven way of dealing with these numerous variables is to dress in layers: As your activity increases, you can shed a layer or two to keep from sweating. Sweating can be dangerous because wet underwear doesn't dry quickly when covered by insulating clothes and leaves you vulnerable to getting chilled. Keeping clothing consistent with your activity level to prevent sweating is an acquired discipline and well worth your attention. My family enjoys teasing me about owning so many kinds of outerwear: My reply is that living at 7,000 feet in the Rockies necessitates having options for temperatures that can range 40 degrees F. in a single day.

B. Roller

Inexpensive canvas chore coat can be washed, an advantage.

If you are cross-country skiing or splitting wood, vests make sense because they don't restrict movement of arms and shoulders. Pockets add utility to any vest. Seek those that button or zip to the neck and have a game pocket in rear, which are practical features. Sweaters are useful because knit wool gives, but they usually don't have pockets. As pullover sweaters wear out consider replacing them with fleece jackets or wool shirts having large pockets. Shirts and jackets have more ways to vent excess heat than pullovers.

A. Golston, The Dome, Boulder Canyon, Boulder, Colorado

Technical rock climbing requires freedom of movement so a sweater works well.

YOUR TOLERANCE

What is your tolerance for cold weather? Those with low blood pressure often feel cold regardless of what they wear, while others wear little more than a t-shirt. You may feel the cold more at the beginning of winter than at the end because your blood is considerably thinner at summer's end. Knowing your tolerance for cold is an essential part of dressing appropriately. Aging usually decreases circulation, which accounts for the migration of elderly to warmer winter climates. The percentage of fat in your body also determines how you feel the cold. If you gain weight throughout your body, as opposed to getting a potbelly, you'll probably need less winter clothing. When in doubt bring extra clothing, especially in the mountains.

Removing sunglasses to use binoculars, author wears synthetic full-zip jacket and water-shedding hat with brim. Also, note hanging black pack straps.

Hot Tips

You can lose a great deal of heat through your head, so hats are a great way to stay warm without adding a lot of bulk. If you need to feel with your fingers to perform a task, wear a pair of fingerless gloves, which keep your fingers warmer than wearing nothing on your hands. If your feet are cold, you're going to be uncomfortable. Buy insulated boots and don't wear steel-toed boots during winter. And despite Hollywood nonsense, experienced outdoorsmen rarely drink alcohol in frigid temperatures.

Outerwear Composition

Never before have outdoorsmen had so many choices for quality cold weather gear. Knowing a garment's intended purpose will enable you to make an intelligent choice. The material that clothes are made of will help to determine their temperature range, resistance to wind, and ability to keep the wearer dry in a downpour. Choices for outerwear include cotton, wool, silk, leather, fur and synthetics.

Cotton has a number of drawbacks for cold weather use: Once wet, it stays wet but does not keep you warm. In fact, it chills you. Exceptions are high-thread-count cottons treated with wax or oil, making them virtually waterproof. Waxed cotton is traditional among British hunters. Australians have their own version called oilskin. The main virtue of these fabrics is their toughness. The drawbacks are they are heavier than laminates and need to be retreated from time to time. Alternatives to wax- and oil-based waterproof treatments also are available and work well.

Laminates (Gore-Tex® was the original engineered, waterproof, and breathable fabric) are in use worldwide. In short, they're a fusion of an outer shell and an inner semipermeable membrane. Manufacturers claim the membrane's micropores are small enough to shed wind-driven rain but large enough to let water vapor escape. For example, the U.S. military uses laminates in its ECWCS, extended cold weather clothing system. Large national online, mail-order vendors like Bass Pro, Cabelas, Eddie Bauer, L.L. Bean, Patagonia, and Mountain Hardware sell a variety of laminates, each claiming to be the best. Most laminates are relatively noisy so if you will be hunting in a laminate make sure it's quiet enough for your purpose. To see a highly technical comparison of laminate "breathability" for several manufacturers visit: verber.com/mark/outdoors/gear/breathability.pdf

Silk is rarely used for outerwear, as it is far less durable than other options. Fur jackets and coats have fallen out of favor for economic and societal reasons: They are extremely expensive and considered ecologically inappropriate. Fur and bulky down garments are too warm for most conditions, but if you will be near either pole, they are proven choices. Using a tough, uninsulated shell to protect thin-skinned down jackets will prolong their life, especially if your work is physically demanding. Down provides superb insulation and is lightweight. If it gets soaked, however, it becomes a sodden clump that cannot be dried satisfactorily without a heat source. If you think you might get down gear wet, choose a synthetic alternative.

Wool has been the fabric of choice of most North American hunters for more than 100 years and with good reason. Few fabrics do so many things so well. It has a wide comfort range and is quieter than many other textiles and keeps the wearer warm after it becomes wet. If your climate is often misty or drizzly, wool is a natural choice: National and state forestry and wildlife agencies provide agents with hard-finish (also called whipcord or twill) wool uniform pants and jackets for winter use. The diagonal ribs add to the fabric's toughness, making them impervious to most thorns. The jackets usually feature four large pockets, which are useful for storing gloves, mittens, hats, or hand-warmers (chemical or fuel-burning). The more time spent in the wilderness, the more you'll appreciate large pockets.

Makers of quality virgin wool outerwear include: C.C. Filson, Woolrich, L.L. Bean, Johnson Woolen Mills, and Bimidji Woolen Mills. (Recycled wool is not an option if you are seeking longevity and durability.) Wool pants can be scratchy so wearing long johns underneath is wise. Cargo pockets greatly enhance usefulness. If you don't want to pay $200 or more for a pair of top quality wool trousers, you can find military surplus pants (many have cargo pockets) for about $25 at surplus stores.

If you work around open flames, untreated synthetics will combust easily and melt, except for Nomex®. Fur is not an option. Leather— worn for generations by aviators, blacksmiths, and welders—is a great choice.

Prices for high-quality garments are rising: Excellent used clothing can be found on eBay and craigslist at a fraction of the original price. I'll choose a quality used hunting jacket (made in North America) over a cheap import any day.

Innerwear Composition

Underwear (also called a base layer) has come a long way since wool was scratchy and synthetics stank after sweating in them. After numerous wash-dry cycles, synthetics can stretch and lose their shape, and wool if washed too hot will shrink considerably. Finely spun merino wool feels as soft as cotton but warms like wool, and newer synthetic underwear is treated with anti-odor chemicals, including silver. After trying numerous synthetics over the years, I switched to merino wool a couple of years ago and haven't regretted it. Silk also is a great choice, especially for fall and spring. Cotton has a wide comfort range, but stay away from it if you plan on doing anything active: it stays damp and clammy long after you've exerted yourself. Nylon is not a suitable fabric for underwear because it refrigerates instead of insulates.

Firearm Tips for Amateurs and Experts

While this portion of the chapter is about firearms, your performance is what counts. A shooter who practices yearly won't be so effective as one who shoots weekly. If you have the luxury of buying another firearm or more ammunition with which to practice, choose the latter.

Your first question might be: What will you ask the firearm to do? Use will dictate the type and features that you'll need. Determine exactly what you require and then do research to discover current values. The rifle you use to take deer is not the one you'll use to bag squirrels. Like many of the topics in this book, the world of firearms is vast and as complicated as you want to make it. For example, you can buy a box of cartridges for your rifle or pistol, or you can choose to buy the components to assemble ammunition: bullets, cartridge cases, primers, powder, as well as automated or manual tools to create finished cartridges.

Consider Used Firearms

If you are familiar with firearms and have good mechanical aptitude, consider buying a used firearm. The first step is finding a reputable firearms dealer with an excellent onsite gunsmith. If you have years of expertise, you can buy a used firearm from a private party. If you're a beginner, buy a new weapon and learn how to handle it safely, shoot it, and maintain it.

What to Look For

Whether handgun or long gun, make sure reliability is your top priority. If the most accurate gun on earth does not fire when needed, it's an expensive club. Finding out which guns are reliable is time well spent. The alternative is to buy unreliable guns that disappoint you instead of making an informed purchase in the first place. If you're new to the world of guns, ask the opinion of several gun dealers (time your visit when business is slow) or friends owning a variety of quality firearms. Most gun enthusiasts are happy to help you. Take notes so you don't forget specifics. You're entering a whole new world.

HANDGUNS

If you're considering a handgun for self-defense, you can choose a double-action revolver (pulling the trigger cocks the hammer) or semi-automatic pistol. Either can be reliable and of high quality.

Single-action revolvers—must be hand cocked each time that you pull the trigger—are quaint, but their day is past. Seek those made in the United States, Austria, Germany, or Switzerland: This isn't prejudice, just experience. Handguns made elsewhere are questionable, because, similar to edged weapons, what you're buying (assuming good design) is the type of steel and how it's heat-treated. If the internal parts wear prematurely, your bargain handgun may become unreliable, so do your research. And look for those made within the last twenty years or so. Trying to find parts for an off-brand or vintage handgun can be frustrating. In general, a semi-auto is a better tool for self-defense than a revolver because it's faster and easier to reload under stress. And if you need to reload fast, you *will* be under stress.

Handgun Caliber

Opinions abound about the optimum handgun load, so there's a wide variety of ammunition. My experience in the U.S. Army plus four decades shooting numerous handguns has taught me that a 9mm NATO, .40 S&W., and .45 ACP (automatic Colt pistol) are excellent choices for autoloaders. I own a .45 ACP. Revolvers chambered in .38 Special, 357 Magnum (a 357 Magnum also will shoot .38 Specials), and .44 Special are proven fight stoppers. Will other cartridges work as well or better? Yes, but where can you buy uncommon ammo easily? Being able to buy ammo in a small-town hardware store is a benefit that may outweigh others. If you find yourself flinching while shooting or reluctant to shoot your handgun regularly, you probably own too much gun. Many shooting ranges rent guns, which is a great way to discover which caliber and handgun fits your hands as well as your needs.

SHOTGUNS

Probably the most versatile of long guns, shotguns can use small shot to bring down birds or can fire rifled slugs to slay deer. Whether you choose a single- or double-barrel break action, bolt action, pump action, or semi-auto action, it will serve a multitude of roles. Shotgun shells come in a variety of loads. Powerful loads are called high brass, and light loads are called low brass, referring to the length of brass on a shell. The amount of gunpowder plus the weight of the shot or slug (similar to a bullet) define the power of a given load. In general the smaller the shot, the less it tends to scatter, so number 8 shot will maintain a tight pattern at a given range while 00 buckshot will spread considerably.

Shotgun Actions

Break actions are the original shotgun configuration and are great for hunting. They're simple and safe because you can carry them broken (action open); plus are great for youths due to their simplicity. Bolt actions are rare and not recommended. Military units and law enforcement agencies have used pump-action scatterguns for decades. An extended magazine enhances your chances in a firefight, but changes the gun's balance point and handling—a worthwhile tradeoff in my opinion. Semi-auto shotguns with extended magazines provide a great deal of firepower for a reasonable price. Do your research and choose the type of action that you are most comfortable with.

Shotgun Gauges

Gauge, like caliber, is a way of expressing barrel diameter. A 12-gauge is larger than a 20-gauge, for example. Choose a gauge that suits your application. Most police forces favor 12-gauge shotguns, and many experienced shot-gunners like the 20-gauge. Either one, loaded with appropriate ammo, will suffice for home defense. In an urban or suburban area, a handgun or rifle bullet that misses an intruder can penetrate drywall plus both sides of your neighbor's apartment or house. A shotgun loaded with smaller shot will not penetrate so well, but will stop an assailant. Magnum shotgun shells contain more gunpowder and shot. While magnum loads extend your reach, they

Coyotes beware. DPMS semi-auto rifle chambered for .308 Winchester with Trijicon 5 X 50 scope. Note bipod protruding from vertical foregrip.

increase recoil, which usually slows your second shot. Standard loads are adequate for home defense.

RIFLES

The rifle's worldwide predominance is mostly due to its long range and accuracy. Armed forces use rifles as the primary weapon for small-unit personnel. Troops use handguns mostly for fighting their way to their rifles. Consider a quality rifle with a detachable box magazine, which facilitates reloads and provides the flexibility to carry different types of ammunition in different magazines. Synthetic stocks are more serviceable and do not expand or contract so much as wood.

Rifle Actions

Bolt actions are traditional for hunting, but semi-autos are gaining acceptance rapidly. Semi-autos are best for defending your loved ones. Manufacturers now make semi-autos that shoot as well or better than some bolt guns. If you can afford only one weapon, consider a reliable semi-auto. Many shooters like pump- and lever actions, but they are not optimum for home defense. Over-and-under combination guns combine a shotgun and rifle barrel. While great for hunting, they are not optimal for defending your castle. However, they are a great deal better than a baseball bat.

Rifle Calibers and Cartridges

Options for rifle loads are numerous, so consider exactly what your cartridge was designed to do. Buying a rifle chambered for killing large game animals will turn small animals into pink mist. No one rifle can do everything well. A rifle chambered for .308 Winchester or 7.62mm X 51mm NATO topped with adequate optics can kill a deer at a quarter mile or more. I use a .308 as an example, because it's a fine all-around rifle, especially in semi-auto. The .223 Remington (5.56mm X 45mm NATO) cartridge has its uses but does not shoot well at distances greater than 400 yards.

Calibers range from cheap-to-shoot .22s to the elephant-stopping .50BMG—the original magnum cartridge, by the way. Choose a common caliber so that you can buy cartridges from a variety of sources. Wildcat (uncommon) cartridges are for hobbyists and tinkerers.

OPTICAL SCOPES AND SIGHTS

A rule of thumb is to spend roughly as much on your optical scope as your weapon. Quality scopes endure heavy recoil year after year, but low-end scopes can break or lose their zero over time. Choose high-end scopes made in the United States, Germany, or Japan. Don't buy a cheap import intending to upgrade: It's a waste of money and reduces your motivation to get what you want. Variable-power optics are more complex (and prone to breakage) than fixed power. Think of a fixed-power scope as a bicycle and a variable-power scope as a motorcycle. Back up your optics with iron sights in case your optics break or lose zero.

If you're shooting at ranges less than 100 yards, consider a reflex sight, which superimposes a dot in the sight's middle. Place the dot on your target and you're ready to fire. Avoid reflex sights (or optical scopes) that use batteries, which can go dead. Look for a reflex sight without batteries. For example, Trijicon makes a reflex sight that uses fiber optics in bright sunlight and a tritium gas capsule for dawn and dusk.

Trijicon RX30 reflex sight is fast—place dot on target and pull trigger—and does not use a battery. DPMS semi-automatic rifle is reliable and accurate.

NIGHT VISION DEVICES

Passive night vision devices (PNVD) are as useful as they are expensive, because they enable you to see and shoot in the dark (not total dark, they need a little light to function). PNVDs are available as goggles or weapon-mounted scopes—both have distinct advantages. For example, you can drive wearing goggles with the vehicle's lights off. Generation 3+ PNVDs are on the market, and their resolution enables users to identify friend or foe instead of merely seeing a grainy, indistinct image. Night operations by NATO troops in Afghanistan and Iraq were successful mostly because they used night vision.

Active night vision uses infrared light to illuminate targets. The problem is that others using night vision can see the source of infrared illumination (you). If you thought finding optimum optics was difficult, wait until you compare these complex pieces of equipment. Buy PNVDs made in the U.S.A.

Thermal-imaging weapon scopes are particularly useful when trying to acquire a target through smoke or fog. If you can afford it—what you'd pay for a new car—by all means look into a thermal-imaging scope. Like any piece of sophisticated electro-optical equipment, you usually get what you pay for so don't consider low-end, thermal-imaging weapon scopes. Spend time doing research on night vision dealers because many are not ethical.

Quality vs. Quantity

Owning several inexpensive firearms of marginal quality is not better than owning one or two of quality. When you're in the field and your weapon jams or breaks, you won't recall its price. Best case, a broken weapon will inconvenience you. Worst case, it can cost your life. Make sure to invest your time in determining exactly what you need, save for it, and don't settle for less. Also set aside some of your budget for ammunition, which you'll use to become proficient with your new acquisition. Firearms for military applications are usually more robust than those made for the occasional hunt: The military version probably is heavier, but I'll choose reliability and heavy-duty construction over saving a few ounces.

Evolution

What was state of the art decades ago may become outdated as gun, ammunition, and optics technology evolve. For example, the Picatinny rail system (MIL-STD-1913) on current military rifles is superior to attaching optics with less robust rings and mounts used on many hunting rifles. Panther Arms (DPMS), Knight's Armament, and Rock River Arms make semi-auto rifles that outperform the venerable M1 and M14 rifles that set the standard for decades. The point is not to buy a gun just to have something. Buy what is current but proven and buy what is reliable and accurate, in that order.

Maintenance

Protect your investment. If you know that you are not diligent about maintenance, consider buying a stainless steel weapon, especially if you're in a high-humidity or marine area. Advances in cleaning products and lubricants have come a long way in the last few years so don't buy the products you've been using for years without checking what's new. Take care of it and it will take care of you.

KNIVES

Knives take many forms—from undersized penknives to huge, over-built knives that male teens lust after. So before buying, decide which tasks you want your new knife to do. For example, fine knives are not pry bars or cleavers or chisels. I know this because in my youth I broke the tip off numerous folding and fixed-blade knives. If you're seeking information about throwing knives or fighting knives, you won't find it here. This portion of the chapter regards knives as cutting tools.

Purpose

If you will carry your knife daily, put considerable thought into what kind to get. For example, if you use your knife often, you may want a folder that clips to your clothing so that you can get to it with one hand, open, use, close, and replace it. If you work in the wilderness, you may want something more substantial, a stout fixed-blade knife for example. Experience dictates that in general a blade longer than six inches is less serviceable. Exceptions include a machete, for clearing vegetation, and a chef's knife, which slices and chops most of lunch and dinner. For daily use, I carry a multi-tool: pliers, serrated blade, plain-edge blade, Phillips screwdriver, and slot-head screwdriver. At least once or twice a day it saves me a trip to the garage for a tool.

Inexpensive vs. Expensive

High-quality knives are expensive, mostly because of the specialized steels that premium knife manufacturers use for blades. Most people don't know how to sharpen premium cutlery, so they give it away or sell it because "it's too hard to sharpen." And they're telling the exact truth. Hard steel takes extra time and effort to sharpen. What many consumers don't consider is that hard steel stays sharp longer so it needs honing less often. A used knife made of excellent steel beats a new knife of marginal steel.

Blade Steel

To make informed cutlery purchases, you need to know which kind of steel (and how it's heat-treated) a manufacturer uses. Steel composition and tempering cannot be explained in a chapter, but a few highlights will provide some direction. The amount of carbon in steel determines its performance characteristics. High-carbon steel is used in most fine blades. Minerals and metals are added to increase toughness and rust resistance, molybdenum or vanadium respectively.

Blades can be cast or forged. In general, forged blades are tougher than cast—the forging process changes the steel's microstructure. Recently, blades have been made from powdered steel, but they are uncommon. Yes, there's a lot to learn about steel, but you need not become a metallurgist to know which brand is best for your requirements. Do enough research to discover a few quality brands that have earned a solid reputation in the highly competitive knife industry and stick with them: For example, Al Mar, Cold Steel, Ka-bar, Randall, SOG, Spyderco, and Victorinox. Buy a knife that you can give proudly to your child or grandchild when the time comes.

Blade Shapes and Sizes

Chisel point, clip point, dagger, drop point, kukri, skinner, Bowie plus hybrids provide buyers with many options. Daggers, sharpened on both edges, are all but useless for most tasks. A practical knife has a strong point, not an overly tapered and thin one. The blade's back or spine can be flat or rounded, which makes splitting wood easier if you need to use a log for a hammer. Saw teeth on the spine are less than ideal when splitting wood or guiding the knife with your weak hand for doing delicate tasks, yet many so-called survival knives have saw-backs. A choil, or unsharpened cutout where handle meets blade, enables your index finger to choke up on the blade providing more control.

Carrying a multi-tool saves author time because he uses it several times a day for small repairs.

LanCay made the M9 bayonet for the U.S. Army. Oval hole in blade fits over stud on sheath to form a heavy-duty scissor for wire fences and the like.

So-called survival knives are fine if they're light enough to carry or compact enough to toss in your glove box or luggage. When the music stops, the knife you're carrying becomes your survival knife. Your Swiss Army knife or little lock-back folder may save your bacon because it's with you.

Many knives are available with a serrated or plain edge or a combination. The advantage of a serrated edge is that it effectively increases the blade's cutting surface area. While harder to sharpen, a serrated knife has more cutting "bite" than a plain edge.

Guards (Hilts)

A full guard protects your fingers—especially when wet—from slipping onto the blade, but so does a half guard, which facilitates placing your thumb on the back of the blade for more control. Some blades have crosscuts on the first inch of the back to prevent your thumb from slipping.

Handle Shapes and Utility

Handle shape is mostly a matter of choice, but look for smooth shapes that you can use for hours. In the dark you need to know instantly which way your blade is oriented: A round handle prevents this. A strong fixed-blade design features a full tang— blade stock extends through the handle. If you must use the pommel (handle end) as a make-do hammer, you'll appreciate this feature.

Lanyards

A lanyard belongs on any fixed-blade knife that you must not lose. If necessary drill a hole in the handle large enough to accept nylon cord. Most high-quality folding knives have holes in the handles. Attach a spring-loaded cord-lock to the lanyard to keep it from slipping off your wrist while working near deep water or anywhere else that you can lose it.

Maintenance

Most steels will rust in high humidity or when left wet, even some stainless steels. Steel wool removes rust and a drop of oil will keep your blade free of oxidation and discoloration. Blood will turn most steels dark gray over time. If this bothers you seek a stainless steel model.

So-Called Survival Products

Ads for survival kits abound. What makes one item a survival product while a similar one is not? Some products are useful and practical, but others are a waste of money. How can you tell the difference?

Utility Trumps Marketing

We often pay more for specialty gear when it includes "survival" in the description: It's human nature to want to survive. Let's examine a few products to discover how to cut through the clutter: The items that you carry daily are what you're likely to have with you if the unexpected happens. Carrying a heavy knife is cumbersome so it often ends up stored somewhere. The little pocketknife or multi-tool you usually carry will serve you well unless you have to down trees or cut through an aircraft skin. An item's utility and quality trump marketing claims.

Survival Kits

Survival kits came about when ship's captains and airplane pilots realized that if their craft crashed, they'd need a few items to help them get back to civilization. Currently, these kits contain all manner of supplies, as the manufacturers want the broadest possible market. It's not a manufacturer's fault that the kit is bloated: It's the consumer's fault for buying it. So what makes a survival kit appropriate? Only you can intelligently anticipate likely scenarios. Are you an executive who wears a suit or dress every day? Are you young or old, fit or fat? Do you live in the city or country? The answers to these questions will help to determine what constitutes a useful kit for you.

What You Own vs. What You Know

Inserting "survival" into a search engine generates more than 100 million results, which means you couldn't sample even a fraction of the products during the next decade. So what's next? The answer

Author wetting bandanna in western Colorado to wear for evaporative cooling in summer.

may not be intuitive to you: Reading and research are great survival tools. Many survivors of life-threatening situations later say, "I had only the clothes on my back." So much for tools and equipment. . . . Faith, determination, fitness, knowledge, and the ability to problem solve and improvise are far more valuable than a sack full of equipment.

Next time you're heading into unfamiliar territory, instead of tossing maps and a GPS device into your bag or pack, take time to familiarize yourself with natural barriers—rivers and mountain ranges, for example—so if you become separated from your possessions you at least have an idea which directions are safe for travel. Useful gear isn't always labeled as such. A bandanna and waterproof matches in your pocket are better than the waist-pack full of supplies that you left under your car seat. And mosquito repellent in the tropics can prove more useful than a day's ration of food.

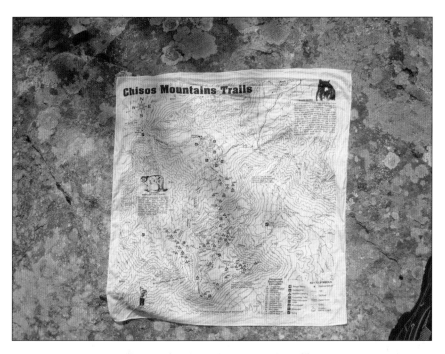

A map printed on a bandanna adds utility.

How to Barter

Ever stand in awe watching a juggler? Skills improve in direct proportion to how often and how long they are practiced. Bartering is no different. People in developing countries barter almost every day. Developed areas like North America use currency and credit for most transactions, so bartering skills are undeveloped. If you want valuable practice, try trading with antique dealers, who are known for their legendary negotiating skills. The following are a few proven tips that you may have forgotten about successful bartering.

Know the Value

Determine the value of the item you're looking for. Depending on the value and importance of what you are seeking, put some extra time into it. And do the same for the item you have to trade. If you are considering trading a handful of DVDs for a few that you haven't viewed, you needn't do much if any research, but if you're about to trade a seldom-used vehicle for a commercial generator, you need to know exactly what you're doing. Without knowing comparable worth, you cannot make an informed transaction. Search engines are a good starting place. For a generator, key words include the make and model. You can insert "review" before the key words to see how others rate the machine. You also may discover that the manufacturer makes diesel and gasoline models, one of which may suit you better. As with any item, condition is critical when determining value. Bring an expert if in doubt.

Establish rapport with your prospective barter partner. This can be a breezy, "Good day, how are you?" A better way of getting on the same wavelength as your prospect is to compliment him or her about something other than the object or service you're seeking to swap. Don't assume anything about your prospect. Asking questions is a great way to find out what motivates your trading partner. By asking questions, you may discover that he or she needs something that you've been meaning to get rid of for years. So questions like:

What else are you looking for? Would you consider swapping for _____? The more you know about your prospect, the better prepared you are to get what you want.

Examine the item you are considering. While you're examining, ask questions about it. You may find that the vehicle you're interested in has been sitting in a garage for more than two years. This inactivity makes the vehicle susceptible to numerous problems. For a fee you can visit carfax.com to learn about a vehicle's history. The vehicle you're considering may have been pulled out of a river.

Trade apples for apples. If both parties to an exchange use an established third-party value for their trade goods, the exchange is usually thought to be fair. For example, if you're exchanging your car for someone's truck, you can both use the Kelley Blue Book (kbb.com) value to establish the worth of each vehicle. The only exception is if the person you want to trade with is a storeowner. Proprietors usually value their merchandise at retail while valuing your item at wholesale. This difference in valuation is partially due to the overhead the proprietor must pay to keep the doors open.

Get creative about the deal and remain flexible about your goal. For example, you've spent six months looking for a suitable used lathe but haven't found one. You find just what you want but the owner wants a bit more than the tool is worth. If you walk away from the deal, you might spend another six months looking. So consider the opportunity cost of walking away. How much money will you save by using the tool for a few months? How much time will you lose by waiting for the "perfect" lathe and price?

Staying Legal

If you are trading children's clothes with your neighbor, the IRS probably doesn't care about you. If you are trading computers, vehicles, or real estate, however, get in touch with a CPA to prevent headaches down the road. Most online barter services (tradeaway. com, u-exchange.com, or barteritonline.com) routinely report their transactions to the IRS. Be aware that business-to-business barter services charge a registration fee plus transaction commissions for their services. These major barter companies facilitate matters by establishing a point or credit system.

Barter companies contend that using credits makes transactions easier and safer—safer because no transfer is made from an account until you authorize the debit. All parties to a transaction register with the company and a transaction does not occur until all participants agree that it's satisfactory. Points and credits provide flexibility. For example, you receive 10,000 trade credits for furniture you no longer need. You can use 2,000 credits to "buy" a computer from one trader, and 4,000 credits for a photocopier from another, and keep the remaining 4,000 credits for future needs. Trade items include aircraft, art, boats, jewelry, livestock, real estate, office equipment, recreational vehicles, and timeshares. Be sure to read barter agreements: You may find that credits expire if not used within a certain time.

INTRUSION DETECTORS AND ALARMS

Advances in perimeter intrusion detectors are announced frequently. These compact, multi-sensing, solid-state devices continue to improve as prices decrease. They can transmit data using radio frequencies or be hardwired. Retail prices range from $150 for something relatively effective to $2,600 or more for a low-voltage, computerized wonder. The most versatile detector and alarm system, however, works *with* sophisticated electronic systems and eats dog food. Consider how dogs can enhance even the most elaborate electronic and human security systems.

As detectors and alarms, dogs have a long and well-documented history. They and their handlers have served with distinction in law enforcement and almost every armed service in the world. They're called guard-, police-, protection-, sentry-, and watchdogs. These hardy working breeds are trainable and depending on temperament

Briards make great alarms because they bark at everything.

and breed, make valuable additions to safeguarding you and your loved ones. Make sure the breed you are considering is suitable for the work you require. The following information comes from ten years' daily experience with a herding breed, plus various mixed-breeds.

Low-Tech Advantages

Unlike the latest electronic intruder-detection system, canines are immune to power outages, coronal mass ejections (solar storms), and electromagnetic pulse caused by nuclear detonations. They do not require back-up power supplies or batteries to function. Their circuitry is naturally robust and will not break if dropped from a few feet onto a hard surface. Immersion in water will not cause short circuits. And perhaps best of all, these detection systems have some of the most acute visual, auditory, and olfactory sensors on the planet. For example, most scent hounds—bloodhounds are legendary—have more than 200 million olfactory receptors while humans have about 5 million. Sight hounds, all seventeen breeds, have the visual acuity to spot small moving objects from hundreds of yards away. The latter are great if your property is fenced. If not, a rabbit may entice a dog to run for miles so that it becomes disoriented and cannot find its way back. They need not be updated or rebooted. And even the best electronic system will not watch your back when you're in the field or keep you warm when camping in cold weather.

Low-Tech Disadvantages

Dogs need access to water and food and require daily walks. They may also sound the alarm when you least want them to. Training a puppy takes time. If you don't have the time or patience to train your dog, there's no shame in taking it to a trainer. Dogs get all kinds of things stuck in their paws, sometimes requiring human intervention. Booties solve this problem and also can prevent snow from melting and freezing between their paw pads. They can become hosts to fleas, which can be remedied with a medicated collar. And they can get *giardia lamblia* by drinking from wild water sources. Dogs chase skunks, which can have terrible consequences if you're sharing a tent with them. They also chase and catch porcupines so keep a multi-tool handy to cut the barbed quills (quills contain air; cutting deflates them) and pull them out with pliers.

High-Tech Advantages

Electronics need not sleep, drink, eat, or be taken for walks. They detect and alert 24–7. They don't become accustomed to neighbors, friends, or certain passers-by: They detect and alert regardless. They need not come inside during storms and blizzards. Electronics do not get worms, pee on the carpet, or wake you at night because they're dreaming.

High-Tech Disadvantages

Electronics can neither smell trespassers nor spot small movements several-hundred yards away. They will not run barking toward a trespasser giving you time to react. Electronic systems often require you to read a lengthy owner's manual, bury seismic sensors, and initialize software.

DOG TRAINING

Socialization is easily done during daily walks—make sure the puppy meets kids, adults, and other animals. Obedience training is a must. Look for a trainer who uses positive reinforcement and trains for the work your dog will perform. Agility training also is useful, particularly if you live in the mountains or a region with a lot of deadfall. Don't ask your dog to do work that it was not bred for. For example, don't ask your Rottweiler to jump and don't ask your bulldog to run.

A Few Breed Characteristics

Some breeds are vigilant by nature. Dobermans are superb athletes and can border on being hyper-vigilant. Rottweilers are remarkably strong and reputed to have the strongest bite strength. While living with two Rotts in Alaska, I was repeatedly surprised by their persistence. I'd throw a twenty-pound log into a deep ravine, and they'd carry it back to me: My arm typically would tire long before they would. They also are loving and protective of family members.

German shepherds are great all-around animals that can be trained to perform several roles. Screen breeders carefully, as German shepherds have been over-bred in the United States and are prone to hip and other health problems. Terriers are full of personality and brave beyond their size. Bernese mountain dogs are athletic working dogs that can do a variety of tasks. Mutts are often a good choice and are always less expensive than purebreds. Small dogs also make effective detectors and alarms: All that's required are keen senses and a bark loud enough to wake or alert you.

What if You Don't like Dogs?

This portion of the chapter regards working breeds, not pets. So whether or not you like dogs, they can patrol large areas less expensively and often more effectively than electronic and human security. And they can do more than detect and alarm: Some breeds will intimidate and deter trespassers and criminals who are seeking targets of opportunity. Imagine you were a burglar considering breaking into a quiet house or a property with two or three barking dogs, which would you choose? If I seem biased, it's because experience has taught me to be so.

What if a Major Disaster Doesn't Occur in Your Lifetime?

Think back on all that you've discovered by becoming more self-reliant and striving to protect your family. You're more knowledgeable, so no matter what the future holds, that knowledge will remain with you. You're more skilled than before you became interested in self-reliance and survival. The combination of knowledge and new skills means you're more self-assured and know your limits better.

The best comparison for becoming more self-reliant—even if a major disaster does not occur—might be training to compete in a sport. You realize that you'll have to change your diet to stay in shape for training—so no more desert or junk foods. You'll have to keep more regular hours and do without beer, so no more pub-crawls. You'll also have to hire a personal trainer and spend more time in the gym. What if you don't make the cut for competing in your sport? You've lost nothing and you've gained by exercising self-discipline, working out on a regular basis, and giving up a few bad habits (as well as possibly losing a few pounds). You have the satisfaction of having tried to accomplish something difficult.

If no disaster comes in your lifetime think of your stockpiled supplies as a retirement plan.

Perhaps Mahatma Gandhi said it best: "Live as if you were to die tomorrow. Learn as if you were to live forever."

RESOURCES

Access to the Internet enables you to discover a world of self-reliance and survival knowledge. While reading and viewing videos showing skills and techniques are important, performing them is critical. For example, tying a knot isn't mastered until *you* repeat it over time so that you have muscle memory. Make hard copies of crucial information so that if the Internet becomes disabled you have what you need.

Gun shows are also a great source of information. While hardware is the main attraction, books and magazines abound. Many small businesses begin by buying a table at a gun show to promote their products and services. And controversial T-shirts and bumper stickers make great gifts. Below are a few core readings and websites that will spark your exploration.

Books

If you read only one book make it Laurence Gonzales' *Deep Survival* (2005), and you are well on your way to understanding the art and science of survival. *Deep Survival* debunks the equipment and physical strength myths. The will to survive, having a strong faith, and being resourceful will help you more than specialized hardware or being athletic.

John "Lofty" Wiseman's *SAS Survival Handbook* (2009) contains abundant, useful information including many black and white as well as color illustrations. The details are what make the book shine. For example, Wiseman doesn't just advise against cutting your only rope; he explains that a knotted rope is 50 percent weaker than a continuous one. Wiseman was a member of Britain's elite Special Air Service for twenty-six years. The book also is available as an e-Book at EbookNetworking.net or you can download the "SAS Guide" as an app (application) for your smart-phone. An *SAS Urban Survival Handbook* (2008) is available from Skyhorse Publishing and can be purchased from any retailer or their website, skyhorsepublishing. com.

James Rawles' *How To Survive the End of the World as We Know It* is a book with all of the fat removed. It provides numerous specifics, including practical products and services. If you discover useful information at survivalblog.com you'll like *How to Survive TEOT-WAWKI*. One glance at the diverse Contents will probably be enough to demonstrate the book's value.

Matthew Stein's *When Technology Fails* (2008) is a massive, large-format book that touches on numerous aspects of self-reliance, survival, and unrelated topics. The book's breadth of scope is breathtaking. While it has a specific social and political point of view, the contents are certainly worthwhile.

The following are books that remove the romance from survival experiences. They inspire by demonstrating the determination that survivors possess.

John Krakauer's *Into Thin Air* (1998) is a gripping account of survival during the controversial 1996 Everest expedition that ended in death for eight climbers, some of whom made gross errors in judgment. Krakauer conveys the difficulty of thinking clearly above 20,000 feet without oxygen. The book and author received flak from accomplished Russian mountaineer Anatoli Boukreev, who died in a Christmas-day avalanche on Annapurna in Nepal in 1997.

Lone Survivor by Marcus Lutrell (2007) tells the bloody tale of four members of U.S. Navy SEAL Team 10 on the mountainous Afghanistan-Pakistan border. When the shooting stopped only the author was alive, but barely. How others saved him and how he made it out to safety has become legendary.

The Rivers Ran East by Leonard Clark (1953) stands the test of time. If you like adventures in the Amazon rainforest, you'll love this book. Clark launched his adventure in 1947 with little money. Sadly, the book is out of print, but you can find a used paperback for about $25 on, no pun intended, Amazon.com.

The Long Walk by Slavomir Rawicz (1956) has become a classic of escape from a Soviet labor camp in Siberia. Without map or compass the author and six fellow prisoners brave hardships with only an axe head and homemade knife. The book illustrates the power of determination in reaching a goal.

We Die Alone by David Howarth (1955) is another classic that takes place during World War II. The author endures the unendurable in Scandinavia and lives to tell about it. Frostbitten and snow-blind, the author soldiers on. It's available as a paperback or electronically as an e-Book.

Online

SurvivalBlog.com opens an online world to you. While many survival and self-reliance sites exist, they do not possess SurvivalBlog's invaluable interactive archive: A free searchable storehouse of practical and proven knowledge from thousands of contributors. A CD of SurvivalBlog archives from 2005 through 2011 is available, which I highly recommend. Founder Jim Rawles, a former U.S. Army intelligence officer, describes his creation as a virtual community. Rawles is the author of best-selling fiction and non-fiction books and has an encyclopedic understanding of survival. He has helped to bring survival and preparation out of the shadows. SurvivalBlog is a one-stop clearinghouse for news articles worldwide, from headlines to obscure sources, which are available by clicking links. Some of these articles scoop newspapers and cable news programs.

AIMPRO Tactical (aimprotactical.com) is the law enforcement service center for Mossberg shotguns. AIMPRO modifies a number of Mossberg pump and semi-automatic tactical shotguns (plus revolvers and semi-automatic pistols) to make them more reliable and better suited to home defense. Founder Michael Shain, former L.A. law enforcement, provides tactical training for law enforcement. Recently, tactical classes have been offered to civilians.

Kifaru International (kifaru.net) makes tough, high-quality packs and tents in the U.S.A. The founder is a master outdoorsman and designer. If you call the Colorado-based company on weekdays a human being answers the phone. The author and Kifaru's founder are planning to make and market survival products in the near future.

Being prepared for what's out there is important—you have to know what to do when everything falls apart. Knowing how to survive the most dangerous places on Earth will prepare you for anything and everything that could possibly go wrong. From packing the proper survival kit, to surviving on the battlefield, being physically fit, and coping in the event of a socio-economic collapse, Soldier of Fortune magazine, along with H. Smith, will make sure that you're never caught off-guard.

The purpose of this book is to provide the reader with real-world, practical information that will help them to not only survive, but thrive during a period that is likely not just another downturn in the economic cycle, but according the many experts, instead the beginning of a long downward slide, and possibly the very peak in our 10,000-year experiment of civilization

Smith will give you the training and knowledge that goes into surviving every dangerous situation imaginable. While you may not plan on being in a war zone, you never know what will happen, so the best thing to always do is be prepared. Learn how to barter and haggle, get the proper camouflage, and choose the right weapon for any situation. Be prepared, be smart, and be able to survive the most dangerous places on Earth.

$14.95 Paperback • ISBN 978-1-62087-098-3